企业级卓越人才培养解决方案"十三五"规划教材

Android 项目式案例开发

天津滨海迅腾科技集团有限公司　主编

南开大学出版社
天　津

图书在版编目 (CIP) 数据

Android 项目式案例开发 / 天津滨海迅腾科技集团有限公司主编 . — 天津：南开大学出版社，2018.8
ISBN 978-7-310-05641-5

Ⅰ. ①A… Ⅱ. ①天… Ⅲ. ①移动终端－应用程序－程序设计 Ⅳ. ①TN929.53

中国版本图书馆 CIP 数据核字 (2018) 第 186880 号

主　编　王　江　刘引涛　刘　涛　陈怀玉
副主编　雷　莹　张在职　牛　芸　顾长青　岳妍妍

版权所有　侵权必究

南开大学出版社出版发行
出版人：刘运峰
地址：天津市南开区卫津路 94 号　邮政编码：300071
营销部电话：(022)23508339　23500755
营销部传真：(022)23508542　邮购部电话：(022)23502200
*
唐山鼎瑞印刷有限公司印刷
全国各地新华书店经销
*
2018 年 8 月第 1 版　　2018 年 8 月第 1 次印刷
260×185 毫米　16 开本　19.5 印张　452 千字
定价：69.00 元

如遇图书印装质量问题，请与本社营销部联系调换，电话：(022)23507125

企业级卓越人才培养解决方案"十三五"规划教材编写委员会

指导专家：周凤华　教育部职业技术教育中心研究所
　　　　　　李　伟　中国科学院计算技术研究所
　　　　　　张齐勋　北京大学
　　　　　　朱耀庭　南开大学
　　　　　　潘海生　天津大学
　　　　　　董永峰　河北工业大学
　　　　　　邓　蓓　天津中德应用技术大学
　　　　　　许世杰　中国职业技术教育网
　　　　　　郭红旗　天津软件行业协会
　　　　　　周　鹏　天津市工业和信息化委员会教育中心
　　　　　　邵荣强　天津滨海迅腾科技集团有限公司
主任委员：王新强　天津中德应用技术大学
副主任委员：张景强　天津职业大学
　　　　　　宋国庆　天津电子信息职业技术学院
　　　　　　闫　坤　天津机电职业技术学院
　　　　　　刘　胜　天津城市职业学院
　　　　　　郭社军　河北交通职业技术学院
　　　　　　刘少坤　河北工业职业技术学院
　　　　　　麻士琦　衡水职业技术学院
　　　　　　尹立云　宣化科技职业学院
　　　　　　王　江　唐山职业技术学院
　　　　　　廉新宇　唐山工业职业技术学院
　　　　　　张　捷　唐山科技职业技术学院
　　　　　　杜树宇　山东铝业职业学院
　　　　　　张　晖　山东药品食品职业学院
　　　　　　梁菊红　山东轻工职业学院
　　　　　　赵红军　山东工业职业学院
　　　　　　祝瑞玲　山东传媒职业学院

王建国	烟台黄金职业学院
陈章侠	德州职业技术学院
郑开阳	枣庄职业学院
张洪忠	临沂职业学院
常中华	青岛职业技术学院
刘月红	晋中职业技术学院
赵　娟	山西旅游职业学院
陈　炯	山西职业技术学院
陈怀玉	山西经贸职业学院
范文涵	山西财贸职业技术学院
任利成	山西轻工职业技术学院
郭长庚	许昌职业技术学院
李庶泉	周口职业技术学院
许国强	湖南有色金属职业技术学院
孙　刚	南京信息职业技术学院
夏东盛	陕西工业职业技术学院
张雅珍	陕西工商职业学院
王国强	甘肃交通职业技术学院
周仲文	四川广播电视大学
杨志超	四川华新现代职业学院
董新民	安徽国际商务职业学院
谭维奇	安庆职业技术学院
张　燕	南开大学出版社

企业级卓越人才培养解决方案简介

企业级卓越人才培养解决方案(以下简称"解决方案")是面向我国职业教育量身定制的应用型、技术技能人才培养解决方案。以教育部—滨海迅腾科技集团产学合作协同育人项目为依托,依靠集团研发实力,联合国内职业教育领域相关政策研究机构、行业、企业、职业院校共同研究与实践的科研成果。本解决方案坚持"创新校企融合协同育人,推进校企合作模式改革"的宗旨,消化吸收德国"双元制"应用型人才培养模式,深入践行基于工作过程"项目化"及"系统化"的教学方法,设立工程实践创新培养的企业化培养解决方案。在服务国家战略:京津冀教育协同发展、中国制造2025(工业信息化)等领域培养不同层次的技术技能人才,为推进我国实现教育现代化发挥积极作用。

该解决方案由"初、中、高"三个培养阶段构成,包含技术技能培养体系(人才培养方案、专业教程、课程标准、标准课程包、企业项目包、考评体系、认证体系、社会服务及师资培训)、教学管理体系、就业管理体系、创新创业体系等;采用校企融合、产学融合、师资融合的"三融合"模式,在高校内共建大数据(AI)学院、互联网学院、软件学院、电子商务学院、设计学院、智慧物流学院、智能制造学院等;并以"卓越工程师培养计划"项目的形式推行,将企业人才需求标准、工作流程、研发规范、考评体系、企业管理体系引进课堂,充分发挥校企双方优势,推动校企、校际合作,促进区域优质资源共建共享,实现卓越人才培养目标,达到企业人才招录的标准。本解决方案已在全国几十所高校开始实施,目前已形成企业、高校、学生三方共赢的格局。

天津滨海迅腾科技集团有限公司创建于2004年,是以IT产业为主导的高科技企业集团。集团业务范围已覆盖信息化集成、软件研发、职业教育、电子商务、互联网服务、生物科技、健康产业、日化产业等。集团以科技产业为背景,与高校共同开展"三融合"的校企合作混合所有制项目。多年来,集团打造了以博士、硕士、企业一线工程师为主导的科研及教学团队,培养了大批互联网行业应用型技术人才。集团先后荣获天津市"五一"劳动奖状先进集体、天津市政府授予"AAA"级劳动关系和谐企业、天津市"文明单位""工人先锋号""青年文明号""功勋企业""科技小巨人企业""高科技型领军企业"等近百项荣誉。集团将以"中国梦,腾之梦"为指导思想,在2020年实现与100所以上高校合作,形成教育科技生态圈格局,成为产学协同育人的领军企业。2025年形成教育、科技、现代服务业等多领域100%生态链,实现教育科技行业"中国龙"目标。

前　言

　　安卓（Android）系统作为一个应运而生的移动终端操作系统，早期的打电话、发短信、浏览网页等功能已经不能满足人们的需求，今天，Android 系统已经发展至成熟阶段，应用领域也不断扩大。从生活中的数字家庭、远程医疗、交通监控，到工业中的智能监控、移动终端、车载配件等都会出现 Android 的身影。

　　本书以项目一手持端、项目二新闻天下和项目三微聊为基础，实现模块化的教学排列方式，以 Android 技术知识点为项目应用方式展现给读者，使读者读完本书后，对 Android 应用开发具有深入、系统的了解，具备项目开发的能力。

　　本书分为基础篇、提升篇和强化篇三个阶段，每个阶段分为三个模块。一、基础篇：登录注册模块，扫码分析模块，扫描记录模块；二、提升篇：新闻阅读模块，图片浏览模块，空气检测模块；三、强化篇：系统及个人模块，好友及群组模块，会话列表模块。循序渐进地讲述 Android 项目开发步骤及流程。通过本书的学习，读者可以更加熟练地使用 Android Studio 进行 Android 项目的开发，掌握项目开发流程的要点，设计出稳定高效的 App。

　　本书每个模块都设有学习目标、任务描述、任务技能点详解、任务实现、任务拓展和任务总结。结构条理清晰、内容详细，任务实现与任务拓展可以将所学的理论知识充分的应用到实战中。

　　本书由王江、刘引涛、刘涛、陈怀玉任主编，由雷莹、张在职、牛芸、顾长青、岳妍妍等共同任副主编，刘涛、陈怀玉负责统稿，王江、刘引涛负责全面内容的规划，雷莹、张在职负责整体内容编排。具体分工如下：模块一项目一至项目三由雷莹、张在职编写，王江负责全面规划；模块二项目一至模块三项目一由牛芸、顾长青共同编写，刘引涛负责全面规划，模块三项目二至项目三由岳妍妍编写，王江负责全面规划。

　　本书理论内容简明、扼要，实例操作讲解细致，步骤清晰，操作步骤附有相对应的效果图，便于读者直观、清晰地看到操作效果，牢记书中的操作步骤。使读者在 Android 的学习过程中能够更加顺利，使自身的 Android 开发能力更上一层楼。

天津滨海迅腾科技集团有限公司
技术研发部

目　录

项目一　物料排序手持端 ··· 1

　模块一　登录注册 ·· 1
　　学习目标 ··· 1
　　学习路径 ··· 1
　　任务描述 ··· 2
　　任务技能 ··· 7
　　　技能点一　MD5 加密 ·· 7
　　　技能点二　SQLite 数据库设计 ·· 9
　　　技能点三　视频背景的实现 ·· 10
　　任务实施 ·· 11
　　任务总结 ·· 36
　　课外扩展 ·· 36
　　　技能扩展——Cookie ·· 36
　　英语角 ·· 39
　　任务习题 ·· 39

　模块二　扫码分析 ·· 41
　　学习目标 ·· 41
　　学习路径 ·· 41
　　任务描述 ·· 41
　　任务技能 ·· 43
　　　技能点一　Zxing ·· 43
　　　技能点二　OkHttp ·· 47
　　任务实施 ·· 50
　　任务总结 ·· 75
　　课外扩展 ·· 75
　　　技能扩展——Android 6.0 权限管理 ··································· 75
　　英语角 ·· 77
　　任务习题 ·· 77

　模块三　扫描记录 ·· 78
　　学习目标 ·· 78
　　学习路径 ·· 79
　　任务描述 ·· 79
　　任务技能 ·· 80

　　　　技能点一　历史记录获取流程 ································ 80
　　　任务实施 ··· 83
　　　任务总结 ··· 89
　　　课外扩展 ··· 90
　　　　技能扩展——Volley ··· 90
　　　英语角 ··· 95
　　　任务习题 ··· 95

项目二　新闻天下 ·· 97
　模块一　新闻阅读 ·· 97
　　　学习目标 ··· 97
　　　学习路径 ··· 97
　　　任务描述 ··· 98
　　　任务技能 ·· 100
　　　　技能点一　Toolbar ··· 101
　　　　技能点二　DrawerLayout ································· 103
　　　　技能点三　TabLayout ····································· 106
　　　任务实施 ·· 112
　　　任务总结 ·· 128
　　　课外扩展 ·· 128
　　　　技能扩展——ToolbarLayout ···························· 128
　　　英语角 ·· 132
　　　任务习题 ·· 132
　模块二　图片浏览 ··· 134
　　　学习目标 ·· 134
　　　学习路径 ·· 134
　　　任务描述 ·· 134
　　　任务技能 ·· 135
　　　　技能点一　RecyclerView ································· 135
　　　　技能点二　SwipeRefreshLayout ························ 140
　　　任务实施 ·· 142
　　　任务总结 ·· 149
　　　课外扩展 ·· 150
　　　　技能扩展——CoordinatorLayout ······················· 150
　　　英语角 ·· 157
　　　任务习题 ·· 157
　模块三　天气检测 ··· 158
　　　学习目标 ·· 158
　　　学习路径 ·· 159

任务描述 ··· 159
　　　任务技能 ··· 160
　　　　技能点一　LocationManager ·· 160
　　　任务实施 ··· 163
　　　任务总结 ··· 173
　　　课外扩展 ··· 173
　　　　技能扩展——LocationManage ······································ 173
　　　英语角 ·· 180
　　　任务习题 ··· 180

项目三　微聊 ··· 182

模块一　系统及个人 ·· 182
　　　学习目标 ··· 182
　　　学习路径 ··· 182
　　　任务描述 ··· 183
　　　任务技能 ··· 186
　　　　技能点一　Retrofit ··· 186
　　　　技能点二　MVP ··· 196
　　　任务实施 ··· 200
　　　任务总结 ··· 223
　　　课外扩展 ··· 223
　　　　技能扩展——MVC ··· 223
　　　英语角 ·· 229
　　　任务习题 ··· 230

模块二　好友及群组 ·· 231
　　　学习目标 ··· 231
　　　学习路径 ··· 231
　　　任务描述 ··· 232
　　　任务技能 ··· 235
　　　　技能点一　RxJava ·· 235
　　　任务实施 ··· 239
　　　任务总结 ··· 266
　　　课外扩展 ··· 266
　　　　技能扩展——Lambda ··· 266
　　　英语角 ·· 269
　　　任务习题 ··· 269

模块三　会话列表 ·· 270
　　　学习目标 ··· 270
　　　学习路径 ··· 270

任务描述 ·· 271
任务技能 ·· 274
　　技能点一　融云 SDK ·· 274
任务实施 ·· 283
任务总结 ·· 295
课外扩展 ·· 295
　　技能扩展——ShareSDK ·· 296
英语角 ··· 299
任务习题 ·· 300

项目一　物料排序手持端

模块一　登录注册

通过实现登录注册功能,学习登录/注册功能实现的方法,了解密码的加密处理方式,熟悉数据存储机制,具备登录注册界面以及功能开发的能力。在任务实现过程中:

- 了解 MD5 加密方式。
- 熟悉 SQLite 的储存机制。
- 掌握实现视频背景的技能。
- 具备登录注册模块开发的能力。

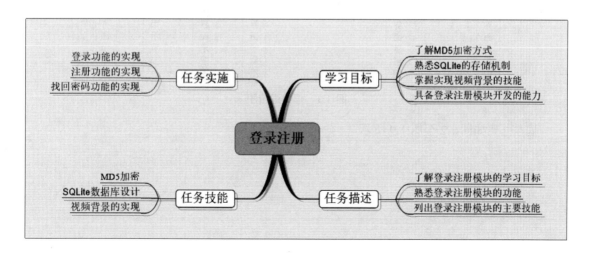

登录/注册的方法有很多,通过手机号进行登录/注册是最常用的方法之一。在物料排序项目中,用户可通过手机号进行登录/注册。考虑到部分用户忘记密码的情况,添加了忘记密码功能,用户在忘记密码的情况下可通过密保验证功能对登录密码进行初始化设置,极大地提高了应用的安全性。

【功能描述】

本模块将实现此项目中的登录注册模块。

- 使用自定义组件对界面进行设置。
- 存储用户名和密码。
- 通过设置密保找回密码。
- 使用 SQLite 数据库存储个人信息。

【基本框架】

本模块共有 9 个界面框架图,各框架图的逻辑关系如图 1.1 所示。

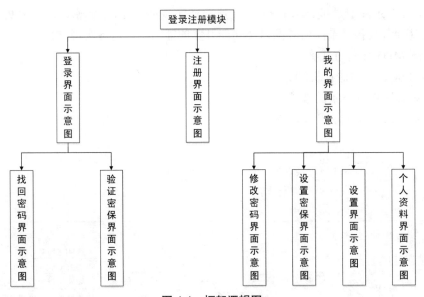

图 1.1　框架逻辑图

基本框架如图 1.2 至图 1.9 所示。

图1.2　登录界面框架图

图1.3　注册界面框架图

图1.4　我的界面框架图

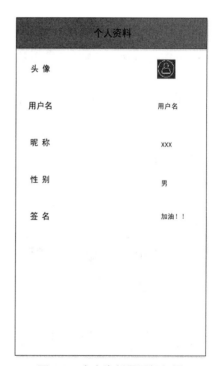

图1.5　个人资料界面框架图

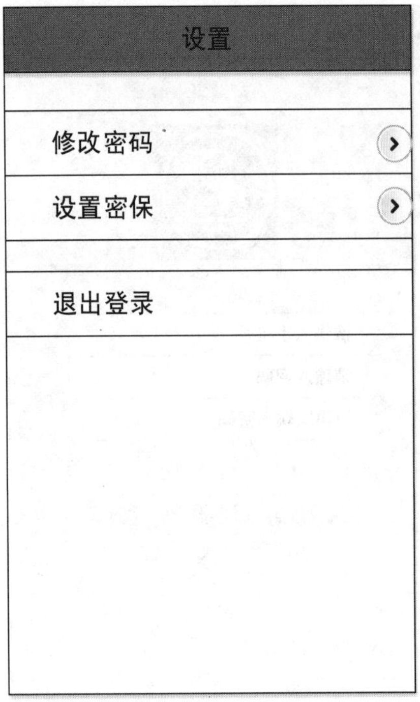

图 1.6　设置界面框架图

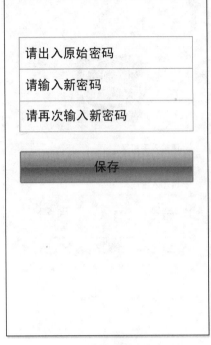

图 1.7　修改密码界面框架图

图 1.8　设置密保界面框架图

图 1.9　找回密码界面框架图

通过本模块的学习,将以上的框架图转换成图 1.10 至图 1.17 所示效果。

项目一 物料排序手持端

图 1.10 登录界面效果图

图 1.11 注册界面效果图

图 1.12 我的界面效果图

图 1.13 个人资料界面效果图

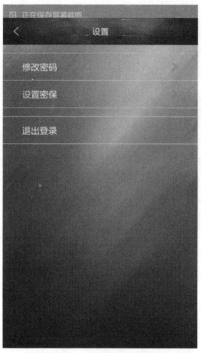

图 1.14 设置界面效果图

图 1.15 修改密码界面效果图

图 1.16 设置密保界面效果图

图 1.17 找回密码界面效果图

技能点一　MD5 加密

如今 MD5 加密技术已经广泛应用到各种领域。因为它的安全可靠,所以被许多开发者应用到了研发过程中,当然本项目在登录验证功能时也采用了 MD5 加密技术,既保证了用户的信息安全,也提高了应用的安全性。

1　MD5 简介

MD5 是 Message digest algorithm(信息摘要算法)的简称,是计算机广泛使用的杂凑算法之一,主流编程语言的加密方式普遍用 MD5 实现。杂凑算法的基础原理是将数据运算为另一固定长度值。MD5 是采用单向的加密算法,其前身有 MD2、MD3 和 MD4,拥有以下两个特性:

- 任意两段明文数据,加密以后的密文是不同的。
- 任意一段明文数据,经过加密以后,其结果必须是不变的。

2　MD5 算法

MD5 以 512 位分组来处理输入信息,且每一分组又被划分为 16 个 32 位子分组,经过一系列的处理后,算法的输出由四个 32 位分组组成,将这四个 32 位分组级联后生成一个 128 位散列值。

3　Android MD5 加密算法

Android 平台支持 java.security.MessageDigest(MD5 加密)包,输入一个 String(需要加密的文本),得到一个加密输出 String(加密后的文本)。具体加密算法代码如 CORE0101 所示:

```
CORE0101    算法签名
String getMD5(String val) throws NoSuchAlgorithmException
package com.tencent.utils;
import java.security.MessageDigest;
import java.security.NoSuchAlgorithmException;
/**
 * 对外提供 getMD5(String) 方法
 * @author randyjia
 **/
```

```
public class MD5 {
    public static String getMD5(String val) throws NoSuchAlgorithmException{
        MessageDigest md5 = MessageDigest.getInstance("MD5");
        md5.update(val.getBytes());
        byte[] m = md5.digest();// 加密
        return getString(m);  }
    private static String getString(byte[] b){
        StringBuffer sb = new StringBuffer();
        for(int i = 0; i < b.length; i ++){
        sb.append(b[i]);
        }
        return sb.toString();
    }  }
```

4 MD5 应用

1. 文件校验

在某些软件下载站点的软件信息中看到其 MD5 值（如图 1.18 所示），它的作用是在下载该软件后，对文件使用专用的软件（如 Windows MD5 Check 等）做一次 MD5 校验，以确保获得的文件与该站点提供的文件为同一文件。利用 MD5 算法来进行文件校验的方案被大量应用到软件下载站、论坛数据库、系统文件安全等方面。

图 1.18 文件校验

2. 登录认证

MD5 广泛用于操作系统的登录认证。当用户登录时，系统将用户输入的密码进行 MD5 Hash 运算，然后再和保存在文件系统中的 MD5 值进行比较，进而确定输入的密码是否正确。通过这样的步骤，系统在并不知道用户密码的情况下即可确定用户登录系统的合法性，避免用户密码被具有系统管理员权限的用户知道。设计微型管理系统亦可如此。

技能点二　SQLite 数据库设计

怎样的存储方式能够达到存储要求去处理数据是开发人员的一道难题。在本项目中，采用 SQLite 数据库作为存储方式，将用户的个人信息全部存储在 SQLite 数据库中，方便查找，更节省空间。

1　SQLite 简介

SQLite 是一款轻量级的关系型数据库，不仅支持标准的 SQL 语句，还遵守 ACID（一个事务本质上有四个特点：Atomicity 原子性、Consistency 一致性、Isolation 隔离性、Durability 耐久性）的关系型数据库管理系统。SQLite 的设计目标是嵌入式的，目前已经在很多嵌入式产品中使用，它占用资源非常低，在嵌入式设备中，可能只需要几百 K 的内存。它能够支持 Windows/Linux/Unix 等主流操作系统，同时能够与很多程序语言相结合，如 Tcl、PHP、Java、C++、Net 等。而 ODBC 接口与 MySQL、PostgreSQL 这两款世界著名的开源数据库管理系统相比，它的处理速度更快。

2　数据库表

本项目在个人资料模块中使用了 SQLite 数据库，表 1.1 是该项目中的数据库表。

表 1.1　该项目的数据库表

字段名称	含义属性	类型	长度	备注
_id	编号	Int	10	PRIMARY KEY（主键） AUTOINCREMENT（自增）
userName	用户名	varchar	50	DEFAULT NULL
nickName	昵称	varchar	50	DEFAULT NULL
sex	性别	varchar	50	DEFAULT NULL
signature	签名	varchar	50	DEFAULT NULL

拓展：在 Android 中，SQLite 是被集成于 Android runtime，每个 Android 应用程序都可以方便的使用 SQLite 数据库。上文列出了在项目中用到的数据库表。想要了解 SQLite 的详细使用方法，扫描右侧二维码一探究竟。

技能点三　视频背景的实现

1　视频背景实现流程

用户在运行项目时,在登录界面会看到一段绚丽的视频作为背景。视频背景的实现步骤如下所示,时序图如图 1.19 所示。

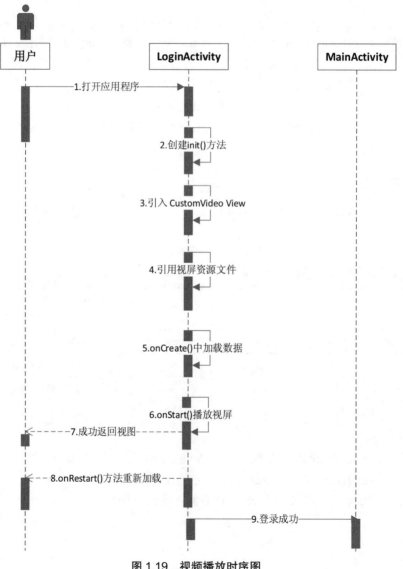

图 1.19　视频播放时序图

- 打开 Android Studio 进入 LoginActivity 主程序。

- 引入播放控件的 VideoView 处理方法 CustomVideoView。
- 布局文件中引用 <kitrobot.com.wechat_bottom_navigation.view.CustomVideoView> 包完成界面设计。
- 将视频文件导入相应资源文件夹下。
- 在主程序 onCreate() 方法中加载视频文件。
- 创建 onStart() 方法开始视频播放。

拓展：通过以上内容，我们了解了本项目中视频背景的实现流程。在 Android 中提供了常见的音频、视频的编码、解码机制。借助于多媒体类 MediaPlayer 的支持，开发人员便于在应用中播放音频、视频。扫描右侧二维码，即可知道 Mediaplayer 的详细使用方法。

通过对以上技能点的学习，下面将实现本项目登录注册模块中一些具体功能。

以下步骤为实现本项目的注册模块，注册功能流程如图 1.20。

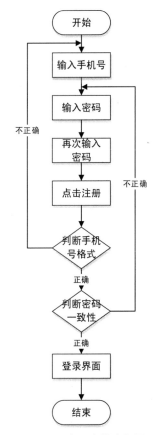

图 1.20　注册功能流程图

第一步：运行项目，首先需要注册用户名。具体代码如 CORE0102 所示。

代码 CORE0102　注册界面实现

```java
public class RegisterActivity extends AppCompatActivity {    // 初始化控件
    private TextView tv_main_title;
    private TextView tv_back;
    private Button btn_register;
    private EditText et_user_name,et_psw,et_psw_again;
    private String userName,psw,pswAgain;
    private RelativeLayout rl_title_bar;
    @Override
    protected void onCreate(Bundle savedInstanceState) {
        super.onCreate(savedInstanceState);
        setContentView(R.layout.activity_register);
        setRequestedOrientation(ActivityInfo.SCREEN_ORIENTATION_PORTRAIT);
        initview();
    }
    private void initview() {
        tv_main_title=(TextView) findViewById(R.id.tv_main_title);
        tv_main_title.setText(" 注册 ");
        tv_back=(TextView) findViewById(R.id.tv_back);
        rl_title_bar=(RelativeLayout) findViewById(R.id.title_bar);
        rl_title_bar.setBackgroundColor(Color.parseColor("#30B4FF"));
        // 从 activity_register.xml 界面布局中获得对应的 UI 控件
        btn_register=(Button) findViewById(R.id.btn_register);
        et_user_name=(EditText) findViewById(R.id.et_user_name);
        et_psw=(EditText) findViewById(R.id.et_psw);
        et_psw_again=(EditText) findViewById(R.id.et_psw_again);
        btn_register.setOnClickListener(new View.OnClickListener() {
            @Override
            public void onClick(View v) {
                // 获取输入在相应控件中的字符串
                getEditString();
                if(isMobileNum(userName)){
                    if(TextUtils.isEmpty(userName)){
                        Toast.makeText(RegisterActivity.this,
" 请输入手机号 ", Toast.LENGTH_SHORT).show();
                        return;
```

```
            }else if(TextUtils.isEmpty(psw)){
                Toast.makeText(RegisterActivity.this,
"请输入密码 ", Toast.LENGTH_SHORT).show();
                return;
            }else if(TextUtils.isEmpty(pswAgain)){
                Toast.makeText(RegisterActivity.this,
"请再次输入密码 ", Toast.LENGTH_SHORT).show();
                return;
            }else if(!psw.equals(pswAgain)){
                Toast.makeText(RegisterActivity.this,
"输入两次的密码不一样 ", Toast.LENGTH_SHORT).show();
                return;
            }else if(isExistUserName(userName)){
                Toast.makeText(RegisterActivity.this,
"此账户名已经存在 ", Toast.LENGTH_SHORT).show();
                return;
            }else{
                Toast.makeText(RegisterActivity.this,
"注册成功 ", Toast.LENGTH_SHORT).show();
                saveRegisterInfo(userName, psw);
                Intent data =new Intent();
                data.putExtra("userName", userName);
                setResult(RESULT_OK, data);
                RegisterActivity.this.finish();
            }} else {
                Toast.makeText(RegisterActivity.this,
"您输入的手机号有误 ", Toast.LENGTH_SHORT).show();  }  });  }
        // 检验电话号码是否正确的方法
    public static boolean isMobileNum(String mobiles) {
        Pattern p = Pattern.compile("^((13[0-9])|(15[^4,\\D])|(18[0,5-9]))\\d{8}$");
        Matcher m = p.matcher(mobiles);
        return m.matches();
    }         }
```

通过完成上述代码，实现注册功能如图 1.21 所示。

图 1.21 注册界面

第二步：若注册成功，则将注册的电话号码与密码存储在"SharedPreferences"内，之后再进行读取，具体代码如 CORE0103 所示。

代码 CORE0103　保存注册信息

```
// 保存账号和密码到 SharedPreferences 中
private void saveRegisterInfo(String userName,String psw){
    String md5Psw= MD5Utils.md5(psw);
    SharedPreferences sp=getSharedPreferences("loginInfo", MODE_PRIVATE);
    SharedPreferences.Editor editor=sp.edit();
    editor.putString(userName, md5Psw);
    editor.commit();
}
    // 读取本文内容
private void getEditString(){
    userName=et_user_name.getText().toString().trim();
    psw=et_psw.getText().toString().trim();
    pswAgain=et_psw_again.getText().toString().trim();
}
    // 判断 SharedPreferences 中是否存在此用户名
private boolean isExistUserName(String userName) {
```

```
boolean has_userName=false;
SharedPreferences sp=getSharedPreferences("loginInfo", MODE_PRIVATE);
String spPsw=sp.getString(userName, "");
if(!TextUtils.isEmpty(spPsw)) {
    has_userName=true;
}
return has_userName;
}
```

以下步骤为实现本项目的登录模块,登录功能流程如图 1.22 所示。

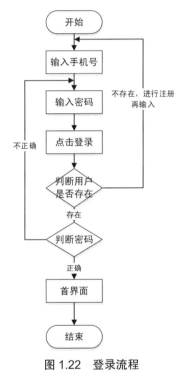

图 1.22 登录流程

第三步:注册好账号,运行成功后首先看到的是登录界面。具体代码如 CORE0104 所示。

代码 CORE0104   登录界面实现
public class LoginActivity extends AppCompatActivity {   // 初始化控件 private TextView tv_register,tv_find_psw; private Button btn_login; private String userName,psw,spPsw; private EditText et_user_name,et_psw; private CustomVideoView videoView;

```java
@Override
protected void onCreate(Bundle savedInstanceState) {
    super.onCreate(savedInstanceState);
    setContentView(R.layout.activity_login);
    setRequestedOrientation(ActivityInfo.SCREEN_ORIENTATION_PORTRAIT);
    // 竖屏显示
    initview();
Init();
}
private void init() {
videoView=(CustomVideoView)findViewById(videoview);
// 使用一段视频作为登录界面的背景
videoView.setVideoURI(Uri.parse("android.resource://"+
getPackageName()+"/"+R.raw.sport));
videoView.start();
// 重复播放视频
videoView.setOnCompletionListener(new MediaPlayer.OnCompletionListener() {
    @Override
    public void onCompletion(MediaPlayer mp) {
        videoView.start();
    }
});
    private void initview() {            // 找到相应的控件
    tv_register=(TextView) findViewById(R.id.tv_register);
    tv_find_psw= (TextView) findViewById(R.id.tv_find_psw);
    btn_login=(Button) findViewById(R.id.btn_login);
    et_user_name=(EditText) findViewById(R.id.et_user_name);
    et_psw=(EditText) findViewById(R.id.et_psw);
    // 从 SharedPreferences 中根据用户名读取密码
    private String readPsw(String userName){
    SharedPreferences sp=getSharedPreferences("loginInfo", MODE_PRIVATE);
    return sp.getString(userName, "");
}    }
```

通过完成上述代码,实现登录界面效果如图 1.23 所示。

项目一 物料排序手持端

图 1.23 登录界面

第四步：判断用户名是否存在，密码是否正确。具体代码如 CORE0105 所示。

代码 CORE0105　判断电话号与密码

```
btn_login.setOnClickListener(new View.OnClickListener() {
    @Override
    public void onClick(View v) {
        userName=et_user_name.getText().toString().trim();
        psw=et_psw.getText().toString().trim();
        String md5Psw= MD5Utils.md5(psw);        // 将密码进行 MD5 加密
        spPsw=readPsw(userName);
        if (isMobileNum(userName)){              // 判断手机号是否正确
            if(TextUtils.isEmpty(userName)){
Toast.makeText(LoginActivity.this, " 请输入手机号 ",
Toast.LENGTH_SHORT).show();
                return;
            }else if(TextUtils.isEmpty(psw)){
Toast.makeText(LoginActivity.this, " 请输入密码 ",
Toast.LENGTH_SHORT).show();
                return;
            }else if(md5Psw.equals(spPsw)){
```

```
                Toast.makeText(LoginActivity.this," 登录成功 ",
Toast.LENGTH_SHORT).show();
                // 保存登录状态
                saveLoginStatus(true, userName);
                Intent data=new Intent();
                data.putExtra("isLogin",true);
                // 将登录成功的信息传入主界面
                setResult(RESULT_OK,data);
                startActivity(new Intent(LoginActivity.this,MainActivity.class));
                LoginActivity.this.finish();
                return;}
else if((spPsw!=null&&!TextUtils.isEmpty(spPsw)&&!md5Psw.equals(spPsw))){
                Toast.makeText(LoginActivity.this," 输入的用户名和密码不一致 ",
Toast.LENGTH_SHORT).show();
                return;
            }else{
                Toast.makeText(LoginActivity.this," 此用户名不存在 ",
Toast.LENGTH_SHORT).show();
            }}
            else {Toast.makeText(LoginActivity.this," 您输入的手机号有误 ",
Toast.LENGTH_SHORT).show();}
        } }); }
```

第五步：保存登录成功的用户名，并将登录成功的信息传入主界面。具体代码如 CORE0106 所示。

代码 CORE0106　保存登录信息传至主界面

```
/**
 * 保存登录状态和登录用户名到 SharedPreferences 中
 */private void saveLoginStatus(boolean status,String userName){
    //loginInfo 表示文件名
    SharedPreferences sp=getSharedPreferences("loginInfo", MODE_PRIVATE);
    SharedPreferences.Editor editor=sp.edit();
    editor.putBoolean("isLogin", status);      // 存入 boolean 类型的登录状态
    editor.putString("loginUserName", userName); // 存入登录状态时的用户名
    editor.commit();   // 提交修改
}
Toast.makeText(LoginActivity.this," 登录成功 ", Toast.LENGTH_SHORT).show();
        // 保存登录状态
```

```
saveLoginStatus(true, userName);
Intent data=new Intent();
data.putExtra("isLogin",true);
// 将登录成功的信息传入主界面
setResult(RESULT_OK,data);
startActivity(new Intent(LoginActivity.this,MainActivity.class));
LoginActivity.this.finish();
```

第六步：登录成功后可跳转至主界面，首先进入"我"的界面，具体代码如 CORE0107 所示。

代码 CORE0107 "我"的界面

```
public class MineFragment extends Fragment {   // 初始化控件
    private  View mView;
    private RelativeLayout title_bar;
    private TextView tv_back,tv_main_title,tv_user_name;
    private ImageView iv_head_icon;
    private LinearLayout ll_head;
    private RelativeLayout rl_course_history,rl_setting;
    @Nullable
    @Override
    // 将布局文件与 Fragment 结合
    public View onCreateView(LayoutInflater inflater, @Nullable ViewGroup container, @Nullable Bundle savedInstanceState) {
        mView=inflater.inflate(R.layout.fragment_mine,null);
        initview();
        return  mView;
    }
    private void initview() {    // 找到布局文件中的所需控件
        title_bar=(RelativeLayout) mView.findViewById(R.id.title_bar);
        title_bar.setBackgroundColor(Color.parseColor("#30B4FF"));
        tv_main_title=(TextView)mView.findViewById(R.id.tv_main_title);
        tv_main_title.setText(" 我 ");
        tv_back=(TextView)mView.findViewById(R.id.tv_back);
        tv_back.setVisibility(View.GONE);
        ll_head=(LinearLayout)mView.findViewById(R.id.ll_head);
        rl_course_history=(RelativeLayout) mView.findViewById(R.id.rl_course_history);
```

```java
        rl_setting = (RelativeLayout) mView.findViewById(R.id.rl_setting);
        title_bar=(RelativeLayout) mView.findViewById(R.id.title_bar);
        iv_head_icon=(ImageView)mView.findViewById(R.id.iv_head_icon);
        tv_user_name=(TextView)mView.findViewById(R.id.tv_user_name);
        setLoginParams(readLoginStatus());   // 界面登录时的控件状态
        ll_head.setOnClickListener(new View.OnClickListener() {
            @Override
            public void onClick(View v) {
                // 判断是否已经登录
                if (readLoginStatus()){
                    // 跳转到登录界面
                    Intent intent=new Intent(getActivity(),UserInfoActivity.class);
                    getActivity().startActivity(intent);
                }
                else {
                    // 跳转到个人信息界面
                    Intent intent=new Intent(getActivity(),UserInfoActivity.class);
                    getActivity().startActivityForResult(intent,1);
                }
            } });
        rl_setting.setOnClickListener(new View.OnClickListener() {
            @Override
            public void onClick(View v) {
                if(readLoginStatus()){
                    // 跳转到设置
                    Intent intent=new Intent(getActivity(),SettingActivity.class);
                    getActivity().startActivityForResult(intent,1);
                }else{
                    Toast.makeText(getActivity()," 您还未登录,请先登录 ",Toast.LENGTH_SHORT).show();
                }           });  }
    private void setLoginParams(boolean isLogin) {
        if(isLogin){
            tv_user_name.setText(AnalysisUtils.readLoginUserName(getActivity()));
        }else{
            tv_user_name.setText(" 点击登录 ");
        }  }
    private boolean readLoginStatus() {
```

```
        SharedPreferences    sp=getActivity().getSharedPreferences("loginInfo",  Context.
MODE_PRIVATE);
        boolean isLogin=sp.getBoolean("isLogin", false);
        return isLogin;
    }    }
```

通过完成上述代码,实现登录成功后,点击进入到"我"的界面如图 1.24 所示。

图 1.24　我的界面

第七步:登录成功后,用户名会显示在界面上。点击头像显示用户的初始信息,包括头像、昵称、签名、性别等,用户也可以对其进行修改具体代码如 CORE0108 所示。

代码 CORE0108　个人信息界面

```
public class UserInfoActivity extends AppCompatActivity implements View.OnClickListener {
// 初始化控件
private TextView tv_back;
private TextView tv_main_title;
private TextView tv_nickName, tv_signature, tv_user_name, tv_sex;
private RelativeLayout rl_nickName, rl_sex, rl_signature, rl_title_bar;
private static final int CHANGE_NICKNAME = 1; // 修改昵称的自定义常量
// 修改个性签名的自定义常量
private static final int CHANGE_SIGNATURE = 2;
```

```java
private String spUserName;
@Override
protected void onCreate(Bundle savedInstanceState) {
    super.onCreate(savedInstanceState);
    setContentView(R.layout.activity_user_info);
    // 竖屏显示
    setRequestedOrientation(ActivityInfo.SCREEN_ORIENTATION_PORTRAIT);
    // 从 SharedPreferences 中获取登录时的用户名
    spUserName = AnalysisUtils.readLoginUserName(this);
    initview();
    initData();
    setListener();
}
private void initview() {    // 获取个人信息界面的控件
    tv_back = (TextView) findViewById(R.id.tv_back);
    tv_main_title = (TextView) findViewById(R.id.tv_main_title);
    tv_main_title.setText(" 个人资料 ");
    rl_title_bar = (RelativeLayout) findViewById(R.id.title_bar);
    rl_title_bar.setBackgroundColor(Color.parseColor("#30B4FF"));
    rl_nickName = (RelativeLayout) findViewById(R.id.rl_nickName);
    rl_sex = (RelativeLayout) findViewById(R.id.rl_sex);
    rl_signature = (RelativeLayout) findViewById(R.id.rl_signature);
    tv_nickName = (TextView) findViewById(R.id.tv_nickName);
    tv_user_name = (TextView) findViewById(R.id.tv_user_name);
    tv_sex = (TextView) findViewById(R.id.tv_sex);
    tv_signature = (TextView) findViewById(R.id.tv_signature);
}
// 为界面控件设置值
private void setValue(UserBean bean) {
    tv_nickName.setText(bean.nickName);
    tv_user_name.setText(bean.userName);
    tv_sex.setText(bean.sex);
    tv_signature.setText(bean.signature);
}
// 设置控件的点击监听事件
private void setListener() {
    tv_back.setOnClickListener(this);
    rl_nickName.setOnClickListener(this);
```

```java
        rl_sex.setOnClickListener(this);
        rl_signature.setOnClickListener(this);
}
@Override
public void onClick(View v) {
    switch (v.getId()) {
        case R.id.tv_back:// 返回键的点击事件
            this.finish();
            break;
        case R.id.rl_nickName:// 昵称的点击事件
            // 获取昵称控件上的数据
            String name = tv_nickName.getText().toString();
            Bundle bdName = new Bundle();
            bdName.putString("content", name);// 传递界面上的昵称数据
            bdName.putString("title", " 昵称 ");
            bdName.putInt("flag", 1);//flag 传递 1 时表示是修改昵称
            enterActivityForResult(ChangeUserInfoActivity.class,
            CHANGE_NICKNAME, bdName);// 跳转到个人资料修改界面
            break;
        case R.id.rl_sex:// 性别的点击事件
            String sex = tv_sex.getText().toString();// 获取性别控件上的数据
            sexDialog(sex);
            break;
        case R.id.rl_signature:// 签名的点击事件
            // 获取签名控件上的数据
            String signature = tv_signature.getText().toString();
            Bundle bdSignature = new Bundle();
            bdSignature.putString("content", signature);
            // 传递界面上的签名数据
            bdSignature.putString("title", " 签名 ");
            bdSignature.putInt("flag", 2);//flag 传递 2 时表示是修改签名
            // 跳转到个人资料修改界面
            enterActivityForResult(ChangeUserInfoActivity.class,
                CHANGE_SIGNATURE, bdSignature);
    break;
        default:
            break;
    } } }
```

通过完成上述代码,实现个人信息界面效果如图 1.25 所示。

图 1.25　个人资料界面

第八步:对个人信息的数据进行储存处理,具体代码如 CORE0109 所示。

代码 CORE0109　个人信息修改

```
// 设置性别的弹出框
private void sexDialog(String sex){
    int sexFlag=0;
    if(" 男 ".equals(sex)){
        sexFlag=0;
    }else if(" 女 ".equals(sex)){
        sexFlag=1;
    }
    final String items[]={" 男 "," 女 "};
    AlertDialog.Builder builder=new AlertDialog.Builder(this);
    builder.setTitle(" 性别 ");
    builder.setSingleChoiceItems(items,sexFlag,new DialogInterface.OnClickListener() {
        @Override
        public void onClick(DialogInterface dialog, int which) {
            // 第二个参数是默认选中的那个项
            dialog.dismiss();
```

```java
            Toast.makeText(UserInfoActivity.this,items[which],Toast.LENGTH_SHORT).show();
                setSex(items[which]);
            }
        });
        builder.create().show(); }
        // 更新界面上的性别数据
private void setSex(String sex){
        tv_sex.setText(sex);
        // 更新数据库中的性别字段
    DBUtils.getInstance(UserInfoActivity.this).updateUserInfo("sex",
            sex, spUserName);
}    private String new_info;// 最新数据
@Override
protected void onActivityResult(int requestCode, int resultCode, Intent data) {
    super.onActivityResult(requestCode, resultCode, data);
    switch (requestCode) {
        case CHANGE_NICKNAME:
            // 个人资料修改界面回传过来的昵称数据
            if (data != null) {
                new_info = data.getStringExtra("nickName");
                if (TextUtils.isEmpty(new_info) || new_info == null) {
                    return;
                }
                tv_nickName.setText(new_info);
                // 更新数据库中的昵称字段
                DBUtils.getInstance(UserInfoActivity.this).updateUserInfo(
                        "nickName", new_info, spUserName);
            }
            break;
        case CHANGE_SIGNATURE:
            // 个人资料修改界面回传过来的签名数据
            if (data != null) {
                new_info = data.getStringExtra("signature");
                if (TextUtils.isEmpty(new_info) || new_info == null) {
                    return;            }
                tv_signature.setText(new_info);
                // 更新数据库中的签名字段
                DBUtils.getInstance(UserInfoActivity.this).updateUserInfo(
                        "signature", new_info, spUserName);
```

```
            }      break;  } }
        // 获取回传数据时需使用的跳转方法,
public void enterActivityForResult(Class<?> to, int requestCode, Bundle b) {
    Intent intent = new Intent(this, to);
    intent.putExtras(b);
    startActivityForResult(intent, requestCode);
}
```

第九步:设置界面,具体代码如 CORE0110 所示。

代码 CORE0110　　设置界面

```
public class SettingActivity extends AppCompatActivity {
// 初始化控件
private TextView tv_main_title;
private TextView tv_back;
private RelativeLayout rl_title_bar;
private RelativeLayout rl_modify_psw,rl_security_setting,rl_exit_login;
public static SettingActivity instance=null;
@Override
protected void onCreate(Bundle savedInstanceState) {
    super.onCreate(savedInstanceState);
    setContentView(R.layout.activity_setting);
    setRequestedOrientation(ActivityInfo.SCREEN_ORIENTATION_PORTRAIT);
    // 设置竖屏
    instance=this;
    initview();
}
private void initview() {
    // 获取控件
    tv_main_title=(TextView) findViewById(R.id.tv_main_title);
    tv_main_title.setText(" 设置 ");
    tv_back=(TextView) findViewById(R.id.tv_back);
    rl_title_bar=(RelativeLayout) findViewById(R.id.title_bar);
    rl_title_bar.setBackgroundColor(Color.parseColor("#30B4FF"));
    rl_modify_psw=(RelativeLayout) findViewById(R.id.rl_modify_psw);
    rl_security_setting=(RelativeLayout) findViewById(R.id.rl_security_setting);
    rl_exit_login=(RelativeLayout) findViewById(R.id.rl_exit_login);
    tv_back.setOnClickListener(new View.OnClickListener() {
```

```
        @Override
        public void onClick(View v) {
            SettingActivity.this.finish();
        }    });
// 跳转至修改密码界面
rl_modify_psw.setOnClickListener(new View.OnClickListener() {
        @Override
        public void onClick(View v) {
            Intent intent=new Intent(SettingActivity.this,ModifyPswActivity.class);
            startActivity(intent);
        }    });
// 设置密保的点击事件
rl_security_setting.setOnClickListener(new View.OnClickListener() {
        @Override
        public void onClick(View v) {
            Intent intent=new Intent(SettingActivity.this,FindPswActivity.class);
            intent.putExtra("from", "security");
            startActivity(intent);
        }    });}
```

通过完成上述代码，实现设置界面效果如图 1.26 所示。

图 1.26　设置界面效果图

第十步：防止密码忘记，在登录成功后要设置密保。具体代码如 CORE0111 所示。

代码 CORE0111　设置密保

```java
public class FindPswActivity extends AppCompatActivity {
// 初始化控件
private EditText et_validate_name,et_user_name;
private Button btn_validate;
private TextView tv_main_title;
private TextView tv_back;
private String from;
private TextView tv_reset_psw,tv_user_name;
@Override
protected void onCreate(Bundle savedInstanceState) {
    super.onCreate(savedInstanceState);
    setContentView(R.layout.activity_find_psw);
    setRequestedOrientation(ActivityInfo.SCREEN_ORIENTATION_PORTRAIT);
    // 获取从登录界面和设置界面传递过来的数据
    from=getIntent().getStringExtra("from");
    initview();    }
private void initview() {
    // 获取控件
    tv_main_title=(TextView) findViewById(R.id.tv_main_title);
    tv_back=(TextView) findViewById(R.id.tv_back);
    et_validate_name=(EditText) findViewById(R.id.et_validate_name);
    btn_validate=(Button) findViewById(R.id.btn_validate);
    tv_reset_psw=(TextView) findViewById(R.id.tv_reset_psw);
    et_user_name=(EditText) findViewById(R.id.et_user_name);
    tv_user_name=(TextView) findViewById(R.id.tv_user_name);
    if("security".equals(from)){
        tv_main_title.setText(" 设置密保 ");
    }else{
        tv_main_title.setText(" 找回密码 ");
        tv_user_name.setVisibility(View.VISIBLE);
        et_user_name.setVisibility(View.VISIBLE);
    }
    tv_back.setOnClickListener(new View.OnClickListener() {
        @Override
```

```java
        public void onClick(View v) {
            FindPswActivity.this.finish();
        } });
    btn_validate.setOnClickListener(new View.OnClickListener() {
        @Override
        public void onClick(View v) {
            String validateName=et_validate_name.getText().toString().trim();
            if("security".equals(from))
            // 设置密保
            if(TextUtils.isEmpty(validateName)){
                Toast.makeText(FindPswActivity.this, " 请输入要验证的姓名 ",
                    Toast.LENGTH_SHORT).show();
                return;
            }else{
                Toast.makeText(FindPswActivity.this, " 密保设置成功 ",
                    Toast.LENGTH_SHORT).show();
                // 保存密保到 SharedPreferences
                saveSecurity(validateName);
                FindPswActivity.this.finish();                }
            }else{    private void savePsw(String userName) {
                // 还原至初始密码"123456"并加密处理
                String md5Psw= MD5Utils.md5("123456");
                SharedPreferences sp=getSharedPreferences("loginInfo", MODE_PRIVATE);
                SharedPreferences.Editor editor=sp.edit();
                editor.putString(userName, md5Psw);
                editor.commit();
            }   }
```

通过完成上述代码,实现设置密保界面效果如图 1.27 所示。

当密保设置成功后,在忘记密码的情况下可在登录界面进行密码找回,找回密码流程如图 1.28 所示。

图 1.27 设置密保界面效果图

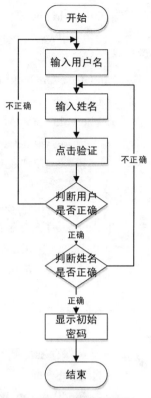

图 1.28 找回密码流程

第十一步：如果忘记密码，可通过密保找回密码。具体代码如 CORE0112 所示。

代码 CORE0112　找回密码
| ```
public class FindPswActivity extends AppCompatActivity {
    // 初始化控件
    private EditText et_validate_name,et_user_name;
    private Button btn_validate;
    private TextView tv_main_title;
    private TextView tv_back;
    private String from;
    private TextView tv_reset_psw,tv_user_name;
    @Override
    protected void onCreate(Bundle savedInstanceState) {
        super.onCreate(savedInstanceState);
        setContentView(R.layout.activity_find_psw);
        setRequestedOrientation(ActivityInfo.SCREEN_ORIENTATION_POR-
TRAIT);
        // 获取从登录界面和设置界面传递过来的数据
``` |
```

```java
            from=getIntent().getStringExtra("from");
"您输入的用户名不存在", Toast.LENGTH_SHORT).show();
                    return;
        initview();
    }
    private void initview() {
        // 获取控件
        tv_main_title=(TextView) findViewById(R.id.tv_main_title);
        tv_back=(TextView) findViewById(R.id.tv_back);
        et_validate_name=(EditText) findViewById(R.id.et_validate_name);
        btn_validate=(Button) findViewById(R.id.btn_validate);
        tv_reset_psw=(TextView) findViewById(R.id.tv_reset_psw);
        et_user_name=(EditText) findViewById(R.id.et_user_name);
        tv_user_name=(TextView) findViewById(R.id.tv_user_name);
            tv_main_title.setText(" 找回密码 ");
        tv_user_name.setVisibility(View.VISIBLE);
        et_user_name.setVisibility(View.VISIBLE);
    }
    tv_back.setOnClickListener(new View.OnClickListener() {
        @Override
        public void onClick(View v) {
            FindPswActivity.this.finish();
        }       });
    btn_validate.setOnClickListener(new View.OnClickListener() {
        @Override
        public void onClick(View v) {
            String validateName=et_validate_name.getText().toString().trim();
            // 通过密保找回密码
            String userName=et_user_name.getText().toString().trim();
            String sp_security=readSecurity(userName);
            if(TextUtils.isEmpty(userName)){
                Toast.makeText(FindPswActivity.this,
"请输入您的用户名", Toast.LENGTH_SHORT).show();
                return;
            }else if(!isExistUserName(userName)){
                Toast.makeText(FindPswActivity.this,
            }else if(TextUtils.isEmpty(validateName)){
```

```java
                        Toast.makeText(FindPswActivity.this,
    " 请输入要验证的姓名 ", Toast.LENGTH_SHORT).show();
                        return;
                    }if(!validateName.equals(sp_security)){
                        Toast.makeText(FindPswActivity.this,
    " 输入的密保不正确 ", Toast.LENGTH_SHORT).show();
                        return;
                    }else{
    private boolean isExistUserName(String userName) {
        boolean hasUserName=false;
        SharedPreferences sp=getSharedPreferences("loginInfo", MODE_PRIVATE);
        String spPsw=sp.getString(userName, "");
        if(!TextUtils.isEmpty(spPsw)) {
            hasUserName=true;
        }
        return hasUserName;
    }
    // 读密保
    private String readSecurity(String userName) {
    SharedPreferences sp=getSharedPreferences("loginInfo", Context.MODE_PRIVATE);
        String security=sp.getString(userName+"_security", "");
        return security;
}
// 存密码对应的密码
    private void saveSecurity(String validateName) {
        SharedPreferences sp=getSharedPreferences("loginInfo", MODE_PRIVATE);
        SharedPreferences.Editor editor=sp.edit();
        editor.putString(AnalysisUtils.readLoginUserName(this)+"_security",
        validateName);
        editor.commit();
    }    }
```

通过完成上述代码，实现找回密码界面效果如图 1.29 所示。

图 1.29 找回密码界面效果图

第十二步：当密码被重置需要重新修改密码。具体代码如 CORE0113 所示。

代码 CORE0113　修改密码

```
public class ModifyPswActivity extends AppCompatActivity {
// 初始化控件
private TextView tv_main_title;
private TextView tv_back;
private EditText et_original_psw,et_new_psw,et_new_psw_again;
private Button btn_save;
private String originalPsw,newPsw,newPswAgain;
private String userName;
@Override
protected void onCreate(Bundle savedInstanceState) {
    super.onCreate(savedInstanceState);
    setContentView(R.layout.activity_modify_psw);
    setRequestedOrientation(ActivityInfo.SCREEN_ORIENTATION_PORTRAIT);
    initview();
    userName= AnalysisUtils.readLoginUserName(this);
}
private void initview() {
```

```java
        // 通过 ID 获取控件
        tv_main_title=(TextView) findViewById(R.id.tv_main_title);
        tv_main_title.setText(" 修改密码 ");
        tv_back=(TextView) findViewById(R.id.tv_back);
        et_original_psw=(EditText) findViewById(R.id.et_original_psw);
        et_new_psw=(EditText) findViewById(R.id.et_new_psw);
        et_new_psw_again=(EditText) findViewById(R.id.et_new_psw_again);
        btn_save=(Button) findViewById(R.id.btn_save);
        tv_back.setOnClickListener(new View.OnClickListener() {
            @Override
            public void onClick(View v) {
                ModifyPswActivity.this.finish();
            }
        });
        btn_save.setOnClickListener(new View.OnClickListener() {
            @Override
            public void onClick(View v) {
                getEditString(); // 读取书写框的文字
                if (TextUtils.isEmpty(originalPsw)) {
                    Toast.makeText(ModifyPswActivity.this,
 " 请输入原始密码 ", Toast.LENGTH_SHORT).show();
                    return;
                } else if (!MD5Utils.md5(originalPsw).equals(readPsw())) {
                    Toast.makeText(ModifyPswActivity.this,
 " 输入的密码与原始密码不一致 ", Toast.LENGTH_SHORT).show();
                    return;
                } else if(MD5Utils.md5(newPsw).equals(readPsw())){
                    Toast.makeText(ModifyPswActivity.this,
 " 输入的新密码与原始密码不能一致 ", Toast.LENGTH_SHORT).show();
                    return;
                } else if (TextUtils.isEmpty(newPsw)) {
                    Toast.makeText(ModifyPswActivity.this,
 " 请输入新密码 ", Toast.LENGTH_SHORT).show();
                    return;
                } else if (TextUtils.isEmpty(newPswAgain)) {
                    Toast.makeText(ModifyPswActivity.this,
 " 请再次输入新密码 ", Toast.LENGTH_SHORT).show();
                    return;
```

```
                } else if (!newPsw.equals(newPswAgain)) {
                    Toast.makeText(ModifyPswActivity.this,
"两次输入的新密码不一致 ", Toast.LENGTH_SHORT).show();
                    return;
                } else {
                    Toast.makeText(ModifyPswActivity.this,
" 新密码设置成功 ", Toast.LENGTH_SHORT).show();
                    // 修改登录成功时保存在 SharedPreferences 中的密码
                    modifyPsw(newPsw);
// 修改成功后跳转至登录界面
Intent intent = new Intent(ModifyPswActivity.this, LoginActivity.class);
                    startActivity(intent);
                    SettingActivity.instance.finish();
                    ModifyPswActivity.this.finish();
                }          }            });    }
// 登录成功后保存密码
private void modifyPsw(String newPsw) {
    String md5Psw= MD5Utils.md5(newPsw);   //MD5 加密
    SharedPreferences sp=getSharedPreferences("loginInfo", MODE_PRIVATE);
    SharedPreferences.Editor editor=sp.edit();
    editor.putString(userName, md5Psw);
    editor.commit(); }
// 读取密码
private String readPsw() {
    SharedPreferences sp=getSharedPreferences("loginInfo", MODE_PRIVATE);
    String spPsw=sp.getString(userName, "");
    return spPsw;}
// 读取书写框中的内容
private void getEditString(){
    originalPsw=et_original_psw.getText().toString().trim();
    newPsw=et_new_psw.getText().toString().trim();
    newPswAgain=et_new_psw_again.getText().toString().trim();
}   }
```

通过完成上述代码，实现效果如图 1.30 和 1.31 所示。

图 1.30　找回密码界面

图 1.31　修改密码界面

本模块介绍了此项目登录注册模块的实现，通过本模块的学习可以了解 SQLite 数据库的设置，掌握 Android 中的数据存储机制，掌握 MD5 在编写过程中的加密方式及原理。学习之后能够实现用户的登录注册功能。

技能扩展——Cookie

1　Cookie 简介

Cookie 由服务器生成，是客户端存储的一种身份凭证。服务器将 Cookie 值发送到客户端/浏览器，客户端将 Cookie 的 key/value 保存到某个目录下，再次请求时，将已得到的 Cookie 值发送到客户端，实现保持登录状态，Cookie 有四个属性如表 1.2 所示：

表 1.2　Cookie 属性

属性	含义
max-age	当 Cookie 生存周期为默认情况时,用户退出应用,Cookie 将会消失
path	表示 Cookie 的所属路径
domain	设置访问域
secure	指定如何传输 Cookie 值

max-age:指定 Cookie 的生存周期(以秒为单位)。默认情况下,Cookie 值只在浏览器的会话期间存在,当用户退出浏览器这些值就会消失。

Path:指定与 Cookie 关联在一起的网页。默认情况下,Cookie 会和创建它的网页、与这个网页处于同一个目录下的网页以及处于该目录下的子目录关联。

domain:设置访问域。如:位于 order.example.com 的服务器要读取 catalog.example.com 设置的 Cookie,这时就要引入 domain 属性。假定由位于 catalog.example.com 的界面创建的 Cookie 把自己的 path 属性设置为 "/",把 domain 属性设置为 ".example.com",那么所有位于 "catalog.example.com" 的网页、位于 "orders.example.com" 的网页以及位于 example.com 域的其他服务器上的网页都能够访问这个 Cookie。如果没有设置 Cookie 的 domain 值,该属性默认值就是创建 Cookie 网页所在服务器的主机名。 注意:不能将 Cookie 的域设置成服务器所在域之外的域。

Secure:指定在网络上如何传输 Cookie 的值。

2　Cookie 的联系与区别

(1)联系

Http 协议是无状态的,后一界面不会记录前一界面的状态,因此可通过 Session 保存上下文信息,从而将客户端的变量保存到服务器中。当客户端访问服务器时,服务器将生成的 Sessionid 发送至客户端并存放到 Cookie 中,当客户端界面提交时,将得到的 Sessionid 返回至服务器,服务器根据 Sessionid 判断用户身份,进而更改对应的客户端信息。因此,Cookie 是 Session 的基础,如果 Cookie 被禁用 Session 也无法使用。

(2)区别

Session 机制用于服务器验证客户端身份,Cookie 与 Session 有以下几点区别:
- Cookie 数据存放在客户端,而 Session 数据存放在服务器。
- Session 较 Cookie 而言较为安全,用户可对存放在本地的 Cookie 进行分析。
- Session 数据在一定时间内存储于服务器上,但当数据增多时,服务器性能就会降低,通过使用 Cookie 可减轻服务器压力。
- Cookie 保存数据量较小,单个不能超过 4K。

3　实现步骤(登录过程)

以下为 Cookie 实现步骤,实现流程如图 1.32 所示。

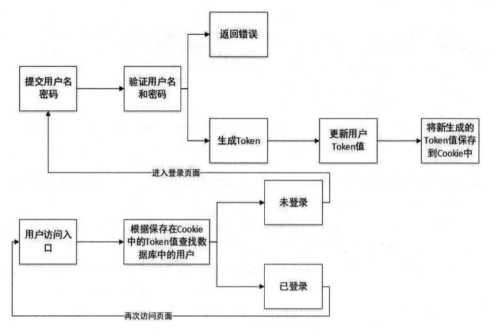

图 1.32 Cookie 实现流程图

第一步：通过 Post 异步请求方法将用户名密码等信息发送至服务器并获取 Cookie 值进行保存，代码如 CORE0114 所示。

代码 CORE0114　　Post 异步请求

```
/** post 异步请求 */
public void postAsynHttp(final String url, final RequestBody formBody, final String key) {
/** 用户登录成功进行数据获取时，需将 Cookie 值发送至服务器 */
Request request = new Request.Builder()
            .url(url)
            .post(formBody)
            .addHeader("cookie", Info.cookie)
            .build();
    Call call = mOkHttpClient.newCall(request);
    call.enqueue(new Callback() {
        @Override
        public void onFailure(Call call, IOException e) {
        }
        @Override
public void onResponse(Call call, Response response) throws IOException {
String phoneString = (String) ShareData.getParam(context, "phoneString", "");
String passwordString=(String)ShareData.getParam(context, "passwordString", "");
```

```
                // 判断数据是否回调成功
                if (response.isSuccessful()) {
                str = response.body().string();
                // 判断得到的数据是否为空
                if (!TextUtils.isEmpty(str)) {
                // 判断当前 url 是否为登录接口
                if (url == Url.login) {
                // 如果是登录接口则获取从服务器传来的 Cookie 值并保存
                List<String> cookieList = response.headers("Set-Cookie");
                String session = cookieList.get(0);
                session = session.substring(0, session.indexOf(";"));
                    Info.cookie = session;
                    }
                Info.map.put(key, str);
                Log.d("str", str + "-----str------");
                    }
                } else {
                Boolean flog = response.isSuccessful();
                Info.map.put(key, flog + "");      }     }     });     }
```

Security	安全	Database	数据库
Video	视频	Register	寄存器
Commit	承诺	Pattern	模式
Matcher	匹配	Bundle	束
Instance	例	Signature	签名

一、选择题

1. 下列 SQLite 叙述不正确的是（　　）。

A. SQLite 是一款轻量级的关系型数据库

B. SQLite 的设计目标是嵌入式的

C. SQLite 只支持 Windows 操作系统

D. SQLite 不仅支持标准的 SQL 语句,还遵守 ACID 的关系型数据库管理系统

2. 关于 MD5 叙述正确的选项是(　　)。

A. MD5 的前身有 MD2 和 MD4

B. MD5 以 514 位分组来处理输入的信息,且每一分组又被划分为 32 个 16 位子分组

C. MD5 的特性之一是:任意一段明文数据,加密以后的密文不能是相同的

D. MD5 的特性之一是:任意一段明文数据,经过加密以后,其结果必须永远是不变的

3. 关于背景视频的实现下列说法正确的是(　　)。

A. 首先引入播放控件的 VideoView 处理方法叫做 CustomVideoView

B. 将视频文件复制到资源文件夹下的 rwa 下

C. 在布局文件中引用这个包名 :"<kitrobot.com.wechat_bottom_navigation.view.CustomVideoViews>"

D. 在主程序 onCreate() 方法中加载数据(播放加载路径,播放,循环播放),在 onRestart() 方法中加载视频

4. 根据任务实施,下列是将登录成功信息传入主界面的方法的是(　　)。

A. data.putExtra("isLogin",true);

B. setResult(RESULT_OK,data);

C. saveLoginStatus(true, userName);

D. editor.putBoolean("isLogin", status);

5. 下列是获取输入控件字符串的方法是(　　)。

A. getText()　　　　　　　　　　B. getEditString();

C. getSharedPreferences()　　　　D. getActivity();

二、填空题

1. MD5 以_____位分组来处理输入的信息,且每一分组又被划分为_____个_____位子分组。

2. MD5 是采用_____加密的加密算法,对于 MD5 而言,有两个特性是很重要的。

3. SQLite 是一款_____的关系型数据库,它不仅支持标准的 SQL 语句,还遵守 ACID 的_____数据库管理系统。

4. ODBC 接口与_____、_____这两款世界著名的开源数据库管理系统相比,它的处理速度_____。

5. 视频背景的实现步骤中首先引入播放控件的_____处理方法叫做 CustomVideoView。

三、上机题

1. 编写代码实现登录功能。

2. 编写代码实现找回密码功能。

模块二　扫码分析

通过扫码以及网络访问的实现,学习二维码/条形码的扫描原理,掌握数据间的传递过程与服务器之间的通讯协议,具备实现扫码数据分析以及网络通信的能力。在任务实现过程中:
- 了解 Zxing 实现二维码/条形码的扫描。
- 熟悉 OkHttp 的使用方法。
- 掌握编写数据传递的技能。
- 具备数据分析网络通信的能力。

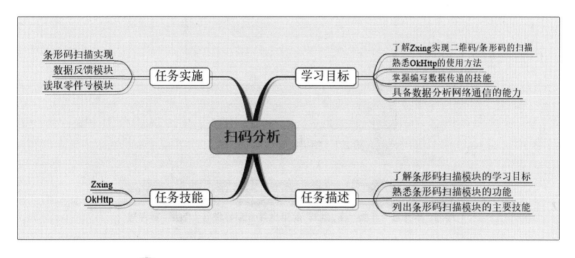

条形码以及二维码的出现极大地方便了人们的生活,在商场购物时,收银员通过扫描条形码进行物品种类、价格的识别,顾客可以通过二维码进行付款。在本项目中,用户可以通过调用手机摄像头进行条形码扫描,返回解码信息并发送到服务器进行数据请求,响应请求后将得到的信息显示到界面中。

【功能描述】

本模块将实现此项目中的条形码扫描模块
- 调用手机摄像头进行条形码扫描。
- 通过 OkHttp 进行数据请求。
- 通过 Intent 进行数据传递。

【基本框架】

基本框架如图 2.1 和图 2.2 所示。

图 2.1　扫描按钮界面框架图

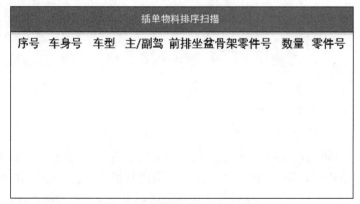

图 2.2　零件列表界面框架图

通过本模块的学习,将以上的框架图转换成图 2.3 和图 2.4 所示效果。

项目一　物料排序手持端

图 2.3　扫描界面

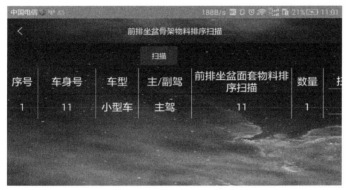

图 2.4　零件号列表界面

技能点一　Zxing

Zxing 是谷歌推出用来识别多种格式条形码的开源项目,本章采用 Zxing 实现扫描二维码和识别二维码两个功能。以下将详细介绍 Zxing 的概念以及使用方法。

1　Zxing 简介

Zxing 是一个开源 Java 类库,用于解析多种格式的 1D/2D 条形码。目标是能够对 QR 编码、Data Matrix、UPC 的 1D 条形码进行解码。其提供了多种平台下的客户端包括:J2ME、J2SE 和 Android。使用 Zxing,可以帮助用户在最短时间内开发出检验 1D/2D 条形码的程序,Zxing 的工作原理是打开手机摄像头,锁定 1D/2D 条形码,并在手机上解析条形码。

最新版本的 Zxing 支持以下编码格式:
- UPC-A and UPC-E。
- EAN-8 and EAN-13。
- Code 39。
- Code 93。
- Code 128。
- QR Code。
- ITF(创新及科技基金)。
- Codabar(库德巴)。
- RSS-14 (all variants——所有的变体)。
- Data Matrix(数据矩阵)。
- PDF 417 ('alpha' quality——'阿尔法'的质量)。
- Aztec ('alpha' quality)。

Zxing 库的主要部分支持以下几个功能:核心代码的使用、适用于 J2SE 客户端的版本、适用于 Android 客户端的版本、Android 的集成等。

2　Zxing 的使用

首先,在(http://code.google.com/p/zxing/)中下载 Zxing 的应用程序包,解压下载程序包后,可以看到整个应用程序主要包含一些组件。如图 2.5 所示。

.github	Add issue template and move supported files to .github
android-core	Manually fix parents of Android libs to 3.3.2-SNAPSHOT
android-integration	Manually fix parents of Android libs to 3.3.2-SNAPSHOT
android	Guard against a few rare errors from Play logs
core	Remove obsolete unrolled loops
docs	Update site for 3.3.1
javase	JAI module isn't needed at compile time
src	Update site version, use HTTPS more in docs
zxing.appspot.com	[maven-release-plugin] prepare for next development iteration
zxingorg	[maven-release-plugin] prepare for next development iteration
.gitattributes	Added .gitattributes to define how git handles the line endings

图 2.5　Zxing 程序包目录

- core：核心包，是整个应用的主要组件组成部分。
- javase：为 PC 端定制的客户端工具。
- android：为 Android 端定制的客户端工具。

1　Zxing 的使用方法：

（1）新建 Android 项目，将 Zxing\zxinglib\src\main\java\com\yzq\zxinglibrary 路径下的所有文件全部复制进入 Wechat-Bottom-navigation\app\src\main\java\kitrobot\com\wechat_bottom_navigation\zxing 的路径下。

（2）将 Zxing\zxinglib\libs\core-3.3.0.jar 文件复制到 Wechat-Bottom-navigation\app\libs 文件夹下，通过导入库类的过程，将 core-3.0.0.jar 包导入项目中。

（3）将 Zxing android 目录下的 res 资源文件拷贝到项目中相应的位置，软件会提示是否覆盖，选择 overwrite all。

（4）将 Zxing\app\src\main\res\values 下的资源文件夹复制进入 Wechat-Bottom-navigation\app\src\main\res\values 文件夹下，并将其中的文件全部替换。如图 2.6 所示。

文件名	日期	类型	大小
colors.xml	2017/10/25 14:41	XML 文档	2 KB
ids.xml	2014/11/12 7:05	XML 文档	1 KB
strings.xml	2017/10/25 14:41	XML 文档	9 KB
styles.xml	2017/10/25 14:43	XML 文档	2 KB

图 2.6　替换文件

（5）替换 CaptureActivity 中的 handleDecode 方法，CaptureActivity 的位置在 com.google.zxing.client.android。具体代码如 CORE0201 所示。

代码 CORE0201　扫描信息，处理返回结果

```java
public void handleDecode(Result rawResult, Bitmap barcode, float scaleFactor) {
    inactivityTimer.onActivity();
    boolean fromLiveScan = barcode != null;
    // 这里处理解码完成后的结果，此处将参数回传到 Activity 处理
    if (fromLiveScan) {
        beepManager.playBeepSoundAndVibrate();
        Toast.makeText(this, " 扫描成功 ", Toast.LENGTH_SHORT).show();
        System.out.println("******result*******"+rawResult.getText().toString());
        Intent intent = new Intent();
        intent.putExtra("codedContent", rawResult.getText().toString());
        intent.putExtra("codedBitmap", barcode);
        setResult(RESULT_OK, intent);
        CaptureActivity.this.finish();
    }
}
```

（6）在 MainActivity 中动态申请权限（Android6.0 之后的要求），添加 onTakePhoto() 方法。具体代码如 CORE0202 所示。

代码 CORE0202　动态获取相机权限

```java
public void onTakePhoto()  {
if (Build.VERSION.SDK_INT>=23)      {
    int request=ContextCompat.checkSelfPermission(activity,
    Manifest.permission.CAMERA);
    if (request!= PackageManager.PERMISSION_GRANTED)
        // 缺少权限，进行权限申请
    {
    ActivityCompat.requestPermissions(activity,new String[]{
        Manifest.permission.CAMERA},123);
        return;     }
    else {
        // 权限同意，不需要处理
        Toast.makeText(activity," 权限同意 ",Toast.LENGTH_SHORT).show();
    }    }
 else{
        // 低于 23 不需要特殊处理
 }    }
```

（7）在调用扫描功能的时候进行界面跳转。具体代码如 CORE0203 所示。

代码 CORE0203　界面跳转

```java
Intent intent = new Intent(activity, CaptureActivity.class);
onTakePhoto();
activity.startActivityForResult(intent,REQUEST_CODE_SCAN);
```

（8）最后将返回值进行回调。具体代码如 CORE0204 所示。

代码 CORE0204　获取返回值

```java
@Override
 protected void onActivityResult(int requestCode, int resultCode, Intent data) {
super.onActivityResult(requestCode, resultCode, data);
       // 扫描二维码 / 条码回传
if (requestCode == REQUEST_CODE_SCAN && resultCode == RESULT_OK) {
    if (data != null) {
        String content = data.getStringExtra(Constant.CODED_CONTENT);
            result.setText(" 扫描结果为 : " + content);
        }    }
```

拓展：想必大家都知道，扫描二维码已经成为现在生活中必不可少的一部分，娱乐消费之后手机一掏"嘀"一声拍屁股走人；出了店门不想走，手机对准旁边的共享单车，不到两秒就成了你的"私人"坐骑。通过上文的学习，相信你们已经能够独立实现扫码功能，那么二维码和条形码的工作原理是什么呢？来，右侧的二维码会告诉你。

技能点二　OkHttp

OKHttp 在接口封装上做得简单易用，和原生的 HttpURLConnection 相比，现已成为 Android 开发者首选的网络通信库。本章采用 OkHttp 作为网络通信库。以下将详细介绍 OkHttp 的概念以及使用方法。

1　OkHttp 简介

OkHttp 是由移动支付 Square 公司开发的一个 Android 网络通信库，能够处理网络请求的开源项目。其用来替代 HttpUrlConnection 和 Apache HttpClient，原因是当前 Android API23 6.0 里面已经删除了 HttpClient。OkHttp 可以处理很多网络疑难杂症，会自动恢复常用的连接问题。如果服务器配置了多个 IP 地址，当第一个 IP 连接失败时，OkHttp 会自动尝试下一个 IP。使用 OkHttp 无需重写程序中的网络代码。OkHttp 实现了几乎和 java.net.HttpURLConnection 一样的 API。如果用了 Apache HttpClient，则 OkHttp 也提供一个对应的 okhttp-apache 模块。

2　OkHttp 的优势

OkHttp 是一款高效的 HTTP 客户端，其具有以下优势：
- 允许连接到同一个主机地址的所有请求，提高请求效率。
- 共享 Socket，减少对服务器的请求次数。
- 通过连接池，减少了请求延迟。
- 缓存响应数据来减少重复的网络请求。
- 减少了对数据流量的消耗。
- 自动处理 GZip 压缩。

3　OkHttp 功能

为了更好的应对网络访问，OkHttp 具有以下功能：
- Get 请求。
- Post 请求。
- 基于 Http 的文件上传。
- 文件下载。
- 加载图片。

- 支持请求回调,直接返回对象、对象集合。
- 支持 session 的保持。

4　OkHttp 使用方式

1. 使用范围
- OkHttp 支持 Android 2.3 及其以上版本。
- 基于 Java,JDK1.7 以上。

2. jar 包准备
在使用的过程中,必须要用到固定的 jar 包,以下是相关 jar 包的下载链接:
- OkHttp:http://square.github.io/okhttp/。
- Okio:http://square.github.io/okio/1.x/okio。

3. 使用教程
向网络发起请求时,最常用的是 GET 和 POST,以下介绍具体使用方法。

（1）GET

OkHttp 中每次网络请求就是一个 Request,Request 里的 url、header 等参数通过 Request 构造出 Call,Call 内部去请求参数得到回复,再将结果告诉调用者,具体代码如 CORE0205 所示。

代码 CORE0205　请求参数

```java
public class TestActivity extends ActionBarActivity {
    private final static String TAG = "TestActivity";
    private final OkHttpClient client = new OkHttpClient();
    @Override
    protected void onCreate(Bundle savedInstanceState) {
        super.onCreate(savedInstanceState);
        setContentView(R.layout.activity_test);
        new Thread(new Runnable() {
            @Override
            public void run() {
                try {
                    execute();
                } catch (Exception e) {
                    e.printStackTrace();
                }
            }
        }).start();
    }
    public void execute() throws Exception {
        Request request = new Request.Builder()
            .url("http://publicobject.com/helloworld.txt")
```

```
        .build();
    Response response = client.newCall(request).execute();
    if(response.isSuccessful()){
        System.out.println(response.code());
        System.out.println(response.body().string());
}}}
```

以上代码通过 Request.Builder 传入 url, 然后由 execute 执行得到 Response, 通过 Response 可以得到 code、message 等信息。这是通过同步的方式去操作网络请求, 而 Android 本身不允许在 UI 线程做网络请求操作, 因此需要开启一个线程。

当然, OKHttp 也支持异步线程并且有回调返回, 有了以上同步的基础, 异步只要稍加改动即可。具体代码如 CORE0206 所示。

代码 CORE0206　回调返回

```
private void enqueue(){
    Request request = new Request.Builder()
        .url("http://publicobject.com/helloworld.txt")
        .build();
    client.newCall(request).enqueue(new Callback() {
        @Override
        public void onFailure(Request request, IOException e) {
        }
        @Override
        public void onResponse(Response response) throws IOException {
            //NOT UI Thread
            if(response.isSuccessful()){
                System.out.println(response.code());
                System.out.println(response.body().string());
}}});}
```

以上代码是在同步的基础上将 execute 改成 enqueue, 并且传入回调接口, 但接口回调的代码是在非 UI 线程上, 所以在有更新 UI 的操作时得用 Handler 或者其他方式。

（2）POST

POST 情况下一般需要传入参数或者一些 Header。

- 传入 Header。具体代码如 CORE0207 所示。

代码 CORE0207　传入 Header

```
Request request = new Request.Builder()
.url("https://api.github.com/repos/square/okhttp/issues")
```

```
.header("User-Agent", "OkHttp Headers.java")
.addHeader("Accept", "application/json; q=0.5")
.addHeader("Accept", "application/vnd.github.v3+json")
.build();
```

传入 POST 参数。具体代码如 CORE0208 所示。

代码 CORE0208 传入 POST

```
RequestBody formBody = new FormEncodingBuilder()
    .add("platform", "android")
    .add("name", "bug")
    .add("subject", "XXXXXXXXXXXXXXX")
    .build();
Request request = new Request.Builder()
    .url(url)
    .post(body)
    .build();
```

由以上代码看出传入 Header 或 POST 参数都是传入 Request 里，因此最后的调用方式和 GET 方式一样。具体代码如 CORE0209 所示。

代码 CORE0209　调用

```
Response response = client.newCall(request).execute();
if (response.isSuccessful()) {
    return response.body().string();
} else {
    throw new IOException("Unexpected code " + response);
}
```

拓展：在日常的开发中，绝大多数应用程序都需要连接网络，发送一些数据给服务端，然后再从服务端获取一些数据。通过以上内容的学习，已经详细了解了 OkHttp 的使用方法。当然在开发过程中可使用的网络框架不只是 OkHttp，扫描右侧二维码查看 Android 中常用的几种网络框架。

通过以上技能点的学习，掌握如何使用 Zxing 实现二维码扫描以及通过 OKHttp 进行网络访问。实现物料排序扫描功能，如图 2.7 所示为实现扫描的具体流程。

项目一 物料排序手持端 51

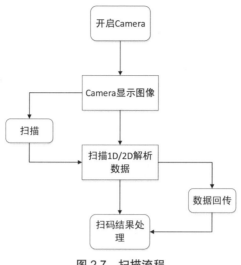

图 2.7 扫描流程

通过以下具体步骤实现扫描功能：

第一步：通过网络 OkHttp 协议，需要添加 okhttp3.2.0.jar 和 okio-1.6.jar 两个 jar 包进行网络的数据传输。具体代码如 CORE0210 所示。

代码 CORE0210　编写网络 okhttp 工具类

```
public class okhttp {
public OkHttpClient mOkHttpClient;
public String str;
// 构建 initOkHttpClient
public void initOkHttpClient(File sdcache) {
    int cacheSize = 10 * 1024 * 1024;
    OkHttpClient.Builder builder = new OkHttpClient.Builder()
            .connectTimeout(15, TimeUnit.SECONDS) // 设置连接的超时时间
            .writeTimeout(20, TimeUnit.SECONDS)   // 设置响应的超时时间
            .readTimeout(20, TimeUnit.SECONDS)    // 请求的超时时间
            .cache(new Cache(sdcache.getAbsoluteFile(), cacheSize));
    mOkHttpClient = builder.build();
}
/**
 * post 请求
 * @param url      请求 url
 * @param callback 请求回调
 * @param formBody 请求参数
 */
```

```java
public void postAsynHttp(final String url, final RequestBody formBody, final String key) {
    Log.d("cookie.url", Info.cookie + "-----str------");
    // 请求数据前的请求体
    Request request = new Request.Builder()
            .url(url)
            .post(formBody)
            .addHeader("cookie", Info.cookie)
            .build();
    Call call = mOkHttpClient.newCall(request);     // 数据回调
    call.enqueue(new Callback() {
        @Override
        public void onFailure(Call call, IOException e) {
            Log.d("e", e + "-------");
        }
        @Override
        public void onResponse(Call call, Response response) throws IOException {
            if (response.isSuccessful()) {
                str = response.body().string();// 把传输数据转化为 string 类型
                if (!TextUtils.isEmpty(str)) {
                    if (url == Url.login) {
                        List<String> cookieList = response.headers("Set-Cookie");
                        // 接收并放入 List 列表中
                        String session = cookieList.get(0);
                        // 截取数据
                        session = session.substring(0, session.indexOf(";"));
                        // 将得到的数据进行截取
                        Info.cookie = session;
                    }
                    Info.map.put(key, str);         // 数据填充
                    Log.d("str", str + "-----str------");}
            } else {
                Log.d("response", response.isSuccessful() + "-------------");
                Boolean flog = response.isSuccessful();
                Info.map.put(key, flog + "");
            } } });  }
```

第二步：绘制扫码框部分，添加 core-3.3.0.jar 包。具体代码如 CORE0211 所示。

代码 CORE0211　绘制扫码框部分

```java
/*
一个位于相机顶部的预览 view，它增加了一个外部部分透明的取景框，以及激光扫描动画和结果组件
*/
public final class ViewfinderView extends View {
    private static final int[] SCANNER_ALPHA = {0, 64, 128, 192, 255, 192, 128, 64};
    private static final long ANIMATION_DELAY = 80L;
    private static final int CURRENT_POINT_OPACITY = 0xA0;
    private static final int MAX_RESULT_POINTS = 20;
    private static final int POINT_SIZE = 6;
    private CameraManager cameraManager;
    private final Paint paint;
    private Bitmap resultBitmap;
    private final int maskColor; // 取景框外的背景颜色
    private final int resultColor;// result Bitmap 的颜色
    private final int laserColor; // 红色扫描线的颜色
    private final int resultPointColor; // 特征点的颜色
    private final int statusColor; // 提示文字颜色
    private int scannerAlpha;
    private List<ResultPoint> possibleResultPoints;
    private List<ResultPoint> lastPossibleResultPoints;
    // 扫描线移动的 y
    private int scanLineTop;
    // 扫描线移动速度
    private final int SCAN_VELOCITY = 10;
    // 扫描线
    Bitmap scanLight;
    public ViewfinderView(Context context, AttributeSet attrs) {
        super(context, attrs);
        paint = new Paint(Paint.ANTI_ALIAS_FLAG);
        Resources resources = getResources();
            // 为扫描框添加背颜色
            maskColor = resources.getColor(R.color.viewfinder_mask);
            resultColor = resources.getColor(R.color.result_view);
            aserColor = resources.getColor(R.color.viewfinder_laser);
            resultPointColor = resources.getColor(R.color.possible_result_points);
```

```
            statusColor = resources.getColor(R.color.status_text);
    scannerAlpha = 0;
    possibleResultPoints = new ArrayList<ResultPoint>(5);
    lastPossibleResultPoints = null;
    scanLight = BitmapFactory.decodeResource(resources,
        R.drawable.scan_light);
}
public void setCameraManager(CameraManager cameraManager) {
    // 扫码成功后震动并发出提示音
    this.cameraManager = cameraManager;
}
@SuppressLint("DrawAllocation")
@Override
public void onDraw(Canvas canvas) {
        if (cameraManager == null) {
// 还没有准备好,若要完成配置还需进行早期的绘图
return;        }
```

第三步:绘制取景框,具体代码如 CORE0212 所示。

代码 CORE0212　　绘制取景框

```
        // frame 为取景框
        Rect frame = cameraManager.getFramingRect();
        Rect previewFrame = cameraManager.getFramingRectInPreview();
        vif (frame == null || previewFrame == null) {
            return;
        }
        int width = canvas.getWidth();
        int height = canvas.getHeight();
        // Draw the exterior (i.e. outside the framing rect) darkened
        // 绘制取景框外的暗灰色的表面,分四个矩形绘制
        paint.setColor(resultBitmap != null ? resultColor : maskColor);
        canvas.drawRect(0, 0, width, frame.top, paint);    // Rect_1
        canvas.drawRect(0, frame.top, frame.left, frame.bottom + 1, paint);//Rect_2
canvas.drawRect(frame.right+1, frame.top, width, frame.bottom + 1, paint); //Rect_3
canvas.drawRect(0, frame.bottom + 1, width, height, paint);    // Rect_4
```

第四步:在扫描矩形上绘制不透明的结果位图,具体代码如 CORE0213 所示。

项目一 物料排序手持端

代码 CORE0213　绘制不透明结果位图

```
// 在扫描矩形上绘制不透明的结果位图
// 如果有二维码结果的 Bitmap，在扫取景框内绘制不透明的 result Bitmap
    paint.setAlpha(CURRENT_POINT_OPACITY);
    canvas.drawBitmap(resultBitmap, null, frame, paint);
} else {
    // 画一个红色"激光扫描仪"线穿过中间显示解码活动
    drawFrameBounds(canvas, frame);
    drawStatusText(canvas, frame, width);
    // 绘制扫描线
    drawScanLight(canvas, frame);
    float scaleX = frame.width() / (float) previewFrame.width();
    float scaleY = frame.height() / (float) previewFrame.height();
    // 绘制扫描线周围的特征点
    List<ResultPoint> currentPossible = possibleResultPoints;
    List<ResultPoint> currentLast = lastPossibleResultPoints;
    int frameLeft = frame.left;
    int frameTop = frame.top;
    if (currentLast != null) {
        paint.setAlpha(CURRENT_POINT_OPACITY / 2);
        paint.setColor(resultPointColor);
        synchronized (currentLast) {
            float radius = POINT_SIZE / 2.0f;
            for (ResultPoint point : currentLast) {
                canvas.drawCircle(frameLeft
                    + (int) (point.getX() * scaleX), frameTop
                    + (int) (point.getY() * scaleY), radius, paint);
            }
        }
    }
postInvalidateDelayed(ANIMATION_DELAY, frame.left - POINT_SIZE,
    frame.top - POINT_SIZE, frame.right + POINT_SIZE,
    frame.bottom + POINT_SIZE);
    }
}
```

第五步：绘制取景框，具体代码如 CORE0214 所示。

代码 CORE0214　绘制取景框

```
/*** 绘制取景框边框 */
private void drawFrameBounds(Canvas canvas, Rect frame) {
```

```
            paint.setColor(Color.WHITE);
            paint.setStrokeWidth(2);
            paint.setStyle(Paint.Style.STROKE);
            canvas.drawRect(frame, paint);
            paint.setColor(Color.BLUE);
            paint.setStyle(Paint.Style.FILL);
            int corWidth = 15;
            int corLength = 45;
            // 左上角
canvas.drawRect(frame.left - corWidth, frame.top, frame.left,
frame.top + corLength, paint);
canvas.drawRect(frame.left - corWidth, frame.top - corWidth,
frame.left+ corLength, frame.top, paint);
        // 右上角
        canvas.drawRect(frame.right, frame.top, frame.right + corWidth,
                frame.top + corLength, paint);
        canvas.drawRect(frame.right - corLength, frame.top - corWidth,
                frame.right + corWidth, frame.top, paint);
        // 左下角
        canvas.drawRect(frame.left - corWidth, frame.bottom - corLength,
                frame.left, frame.bottom, paint);
        canvas.drawRect(frame.left - corWidth, frame.bottom, frame.left
                + corLength, frame.bottom + corWidth, paint);
        // 右下角
        canvas.drawRect(frame.right, frame.bottom - corLength, frame.right
                + corWidth, frame.bottom, paint);
        canvas.drawRect(frame.right - corLength, frame.bottom, frame.right
                + corWidth, frame.bottom + corWidth, paint);
    }
```

第六步：绘制提示文字，具体代码如 CORE0215 所示。

代码 CORE0215　绘制提示文字

```
/* 绘制提示文字 */
private void drawStatusText(Canvas canvas, Rect frame, int width) {
        // 提示文字 1 的内容
        String statusText1 = getResources().getString(
        R.string.viewfinderview_status_text1);
```

```java
            // 提示文字 2 的内容
            String statusText2 = getResources().getString(
                R.string.viewfinderview_status_text2);
        int statusTextSize = 45;
        int statusPaddingTop = 180;
        paint.setColor(statusColor);
        paint.setTextSize(statusTextSize);
        int textWidth1 = (int) paint.measureText(statusText1);  // 提示文字 1 的宽度
            canvas.drawText(statusText1, (width - textWidth1) / 2, frame.top
                - statusPaddingTop, paint);
        int textWidth2 = (int) paint.measureText(statusText2);  // 提示文字 2 的宽度
            canvas.drawText(statusText2, (width - textWidth2) / 2, frame.top
                - statusPaddingTop + 60, paint);
    }
    /*** 绘制移动扫描线 */
    private void drawScanLight(Canvas canvas, Rect frame) {
        if (scanLineTop == 0) {
            scanLineTop = frame.top;
        }
        if (scanLineTop >= frame.bottom) {
            scanLineTop = frame.top;
        } else {
            scanLineTop += SCAN_VELOCITY;
        }
        Rect scanRect = new Rect(frame.left, scanLineTop, frame.right,
                scanLineTop + 30);
        canvas.drawBitmap(scanLight, null, scanRect, paint);
    }
    public void drawViewfinder() {
        Bitmap resultBitmap = this.resultBitmap;
        this.resultBitmap = null;
        if (resultBitmap != null) {
            resultBitmap.recycle();
        }
        invalidate();
    }
```

第七步：扫描二维码界面的编写，实现如图 2.8 所示效果。具体代码如 CORE0216 所示。

代码 CORE0216 二维码扫描功能实现

```java
/*
这个 activity 打开相机,在后台线程做常规的扫描;它绘制了一个结果 view 来帮助
正确地显示条形码,在扫描的时候显示反馈信息,然后在扫描成功的时候覆盖扫描结
果
*/
public final class CaptureActivity extends Activity implements
SurfaceHolder.Callback {
    private static final String TAG = CaptureActivity.class.getSimpleName();
    // 相机控制
    private CameraManager cameraManager;
    private CaptureActivityHandler handler;
    private ViewfinderView viewfinderView;
    private boolean hasSurface;
    private IntentSource source;
    private Collection<BarcodeFormat> decodeFormats;
    private Map<DecodeHintType, ?> decodeHints;
    private String characterSet;
    // 电量控制
    private InactivityTimer inactivityTimer;
    // 声音、震动控制
    private BeepManager beepManager;
    private ImageButton imageButton_back;
    public ViewfinderView getViewfinderView() {
        return viewfinderView;
    }
    public Handler getHandler() {          // 调用 handler 线程
        return handler;
    }
    public CameraManager getCameraManager() {   // 检测设备的系统服务
        return cameraManager;
    }
    public void drawViewfinder() {         // 绘制扫码框
        viewfinderView.drawViewfinder();
    }
```

项目一　物料排序手持端　　59

图 2.8　二维码扫描界面

第八步：照相机的辅助类，具体代码如 CORE0217 所示。

```
代码CORE0217    照相机辅助类
/*
    OnCreate 中初始化一些辅助类，如 InactivityTimer（休眠）、Beep（声音）以及 AmbientLight（闪光灯）
*/
@Override
public void onCreate(Bundle icicle) {
    super.onCreate(icicle);
// 保持 Activity 处于唤醒状态
Window window = getWindow();
window.addFlags(WindowManager.LayoutParams.FLAG_KEEP_SCREEN_ON);
    setContentView(R.layout.capture);
    hasSurface = false;
    inactivityTimer = new InactivityTimer(this);        // 初始化定时器
    beepManager = new BeepManager(this);
    // 在二维码解码成功时播放"bee"的声音，同时还可以震动
    imageButton_back = (ImageButton) findViewById(
    R.id.capture_imageview_back);
    imageButton_back.setOnClickListener(new View.OnClickListener() {
```

```java
            @Override
            public void onClick(View v) {
                finish();        });    }
    @Override
    protected void onResume() {
        super.onResume();
        /*
        CameraManager 必须在这里初始化,而不是在 onCreate() 中。这是因为当第一次进
入时需要显示帮助页,这时并不想打开 Camera,当测量屏幕大小扫描框的尺寸不正确
时会出现 bug
        */
        cameraManager = new CameraManager(getApplication());
        viewfinderView = (ViewfinderView) findViewById(R.id.viewfinder_view);
        viewfinderView.setCameraManager(cameraManager);
        handler = null;
        SurfaceView surfaceView = (SurfaceView) findViewById(R.id.preview_view);
        SurfaceHolder surfaceHolder = surfaceView.getHolder();
        if (hasSurface) {
            // activity 在 paused 时但不会 stopped,因此 surface 仍旧存在;
            // surfaceCreated() 不会调用,因此在这里初始化 camera
            initCamera(surfaceHolder);
        } else {
            // 重置 callback,等待 surfaceCreated() 来初始化 camera
            surfaceHolder.addCallback(this);
        }
        beepManager.updatePrefs();
        inactivityTimer.onResume();
        source = IntentSource.NONE;
        decodeFormats = null;
        characterSet = null;
    }
    @Override
    protected void onPause() {
        if (handler != null) {
            handler.quitSynchronously();
            handler = null;
        }
```

第九步:重新唤起相机定时器,具体代码如 CORE0218 所示。

项目一　物料排序手持端

代码 CORE0218　重新唤起定时器

```
inactivityTimer.onPause();        // 重新唤醒定时器
    beepManager.close();          // 关闭扫码成功后的震动以及提示音
    cameraManager.closeDriver();  // 关闭检测设备的系统服务
    if (!hasSurface) {
    SurfaceView surfaceView = (SurfaceView) findViewById(R.id.preview_view);
    SurfaceHolder surfaceHolder = surfaceView.getHolder();
    surfaceHolder.removeCallback(this);
    }
    super.onPause();
}
@Override
protected void onDestroy() {
    inactivityTimer.shutdown();
    super.onDestroy();
}
@Override
public void surfaceCreated(SurfaceHolder holder) {
    if (!hasSurface) {
        hasSurface = true;
        initCamera(holder);
    } }
@Override
public void surfaceDestroyed(SurfaceHolder holder) {
    hasSurface = false;
}
@Override
public void surfaceChanged(SurfaceHolder holder, int format, int width, int height) {
}
```

第十步：扫描成功后进行消息反馈，具体代码如 CORE0219 所示。

代码 CORE0219　扫描成功后反馈信息

```
/*** 扫描成功,处理反馈信息 **/
public void handleDecode(Result rawResult, Bitmap barcode, float scaleFactor) {
    inactivityTimer.onActivity();
    boolean fromLiveScan = barcode != null;
    // 这里处理解码完成后的结果,此处将参数回传到 Activity 处理
```

```java
        if (fromLiveScan) {
            beepManager.playBeepSoundAndVibrate();
            Toast.makeText(this, " 扫描成功 ", Toast.LENGTH_SHORT).show();
            System.out.println("******result*******"+rawResult.getText().toString());
            Intent intent = new Intent();
            intent.putExtra("codedContent", rawResult.getText().toString());
            intent.putExtra("codedBitmap", barcode);
            setResult(RESULT_OK, intent);
            finish();
        }
    }
/*** 初始化 Camera*/
private void initCamera(SurfaceHolder surfaceHolder) {
    if (surfaceHolder == null) {
        throw new IllegalStateException("No SurfaceHolder provided");
    }
    if (cameraManager.isOpen()) {
        return;
    }
        try {
            // 打开 Camera 硬件设备
            cameraManager.openDriver(surfaceHolder);
            // 创建一个 handler 来打开预览,并抛出一个运行时异常
            if (handler == null) {
                handler = new CaptureActivityHandler(this, decodeFormats,
                        decodeHints, characterSet, cameraManager);
            }
        } catch (IOException ioe) {
            Log.w(TAG, ioe);
            displayFrameworkBugMessageAndExit();
        } catch (RuntimeException e) {
            Log.w(TAG, "Unexpected error initializing camera", e);
            displayFrameworkBugMessageAndExit();
        }
    }
/* 显示底层错误信息并退出应用 */
private void displayFrameworkBugMessageAndExit() {
    AlertDialog.Builder builder = new AlertDialog.Builder(this);
    builder.setTitle(getString(R.string.app_name));
    builder.setMessage(getString(R.string.msg_camera_framework_bug));
```

项目一　物料排序手持端

```
    builder.setPositiveButton(R.string.button_ok, new FinishListener(this));
    builder.setOnCancelListener(new FinishListener(this));
    builder.show();
}   }
```

第十一步：调用扫码功能，对物料信息表进行扫码，这里需要添加 gson-2.7.jar 包。具体代码如 CORE0220 所示。

代码 CORE0220　　调用扫码功能对物料信息表进行扫描

```
public class HomeFragment extends Fragment {
// 初始化控件
private  View mView;
private RelativeLayout title_bar;
private TextView tv_back,tv_main_title;
private ListView lv_list;
private String content,title;
List<Map<String,String>> list = new ArrayList<>();
private Activity activity;
Intent intent = new Intent();              // 界面跳转方法
okhttp okhttp = new okhttp();              // 调用网络接口
private static final String DECODED_CONTENT_KEY = "codedContent";
private static final int REQUEST_CODE_SCAN = 0x0000;
@Nullable
@Override
// 自定义的界面
public View onCreateView(LayoutInflater inflater, @Nullable ViewGroup container, @Nullable Bundle savedInstanceState) {
    mView=inflater.inflate(R.layout.fragment_home,null);
    initview();
    File sdcache = activity.getExternalCacheDir();      // 定义缓存路径
    okhttp.initOkHttpClient(sdcache);
    RequestBody body = new FormBody.Builder().add ("JOSN","Item").build();
//post 请求数据
okhttp.postAsynHttp( "http://192.168.2.103:8080/SSMDemo/users/pro5",
    body,"ItemText");
    handler.postDelayed(new Runnable() {
        @Override
        public void run() {
```

```java
            try {
                // 将数据传入 handler 线程
                String Iknowstr = Info.map.get("ItemText");
                Message msg = new Message();
                msg.what = 0;
                msg.obj =Iknowstr;
                handler.sendMessage(msg);
            }
            catch (Exception e) {
            } }  },500);
    return mView;
}
```

第十二步：Handler 线程操作，将获取到的数据进行拆分，实现如图 2.9 所示效果。具体代码如 CORE0221 所示。

代码 CORE0221　　Handler 线程操作

```java
//Handler 线程操作
Handler handler = new Handler(){
    @Override
    public void handleMessage(Message msg) {
        switch (msg.what){
            case 0:
                list.clear();
                String Items = (String) msg.obj;
                System.out.printf("----Items---" + Items);
                try {
                    // 将获取到的数据进行拆分
                    JSONObject object = new JSONObject(Items);
                    JSONArray array = object.getJSONArray("text");
                    for (int i=0;i<array.length();i++){
                Map<String,String> map = new HashMap<String, String>();
                        map.put("text",array.getString(i));
                        list.add(map);
                    }
                    // 适配器填充
                    ListAdapter adapter = new ListAdapter(getActivity(),list);
                    lv_list.setAdapter(adapter);
```

```
                    adapter.notifyDataSetChanged();
                }
                catch (Exception e){};
                break;
            case 1:
                String ItemNames = (String) msg.obj;
                try {
                    JSONObject object = new JSONObject(ItemNames);
                    JSONArray array = object.getJSONArray("item_name");
                    for (int i=0;i<array.length();i++){
                        title=array.getString(i);
                    }   intent.putExtra("title",title);
                    intent.setClass(activity, OneActivity.class);
                    activity.startActivity(intent);                        }
                catch (Exception e){};
                break;
        }           }          };
private void initview() {
    // 通过 mView 获取控件
    title_bar=(RelativeLayout) mView.findViewById(R.id.title_bar);
    title_bar.setBackgroundColor(Color.parseColor("#30B4FF"));
    tv_main_title=(TextView)mView.findViewById(R.id.tv_main_title);
    tv_main_title.setText(" 扫描 ");
    tv_back=(TextView)mView.findViewById(R.id.tv_back);
    tv_back.setVisibility(View.GONE);
    lv_list=(ListView) mView.findViewById(R.id.lv_list);          }
```

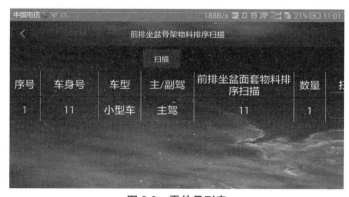

图 2.9　零件号列表

第十三步：动态获取照相机权限，具体代码如 CORE0222 所示。

代码 CORE0222　　动态获取照相机权限

```
// 动态获取相机权限
public void onTakePhoto()  {
    if (Build.VERSION.SDK_INT>=23)      {
int request=ContextCompat.checkSelfPermission(activity,
        Manifest.permission.CAMERA);
        if (request!= PackageManager.PERMISSION_GRANTED) {
        // 缺少权限，进行权限申请
            ActivityCompat.requestPermissions(activity,new String[]
{Manifest.permission.CAMERA},123);
            return;
    }
        else  {
            // 权限同意，不需要处理
            Toast.makeText(activity," 权限同意 ",Toast.LENGTH_SHORT).show();
    }  }
        else{
            // 低于 23 不需要特殊处理
    }  }
// 扫描结果回调
@Override
public void onActivityResult(int requestCode, int resultCode, Intent data) {
    super.onActivityResult(requestCode, resultCode, data);
        if (data != null) {
            content = data.getStringExtra("codedContent");
    }  }
@Override
// 将 fragment 转为 Activity
public void onAttach(Activity activity) {
    super.onAttach(activity);
    this.activity = activity;     }  }
```

第十四步：点击列表进行界面跳转，具体代码如 CORE0223 所示。

代码 CORE0223　　点击列表进行跳转，并携带数据

```
lv_list.setOnItemClickListener(new AdapterView.OnItemClickListener() {
        @Override
    public void onItemClick(AdapterView<?> parent, View view, int position, long id) {
```

```
intent = new Intent(activity, CaptureActivity.class);
// 界面跳转
onTakePhoto();                    }
activity.startActivity(intent);
switch (position){
    case 0:
        onTakePhoto();
        intent.putExtra("title"," 前排坐盆面套物料排序扫描 ");
        intent.putExtra("url","");
        intent.setClass(activity, OneActivity.class);
        activity.startActivity(intent);
        break;
    case 1:
        onTakePhoto();
        intent.putExtra("title"," 前排靠背面套物料排序扫描 ");
        intent.putExtra("url","");
        intent.setClass(activity, OneActivity.class);
        activity.startActivity(intent);
        break;
    case 2:
        onTakePhoto();
        intent.putExtra("title"," 前排坐盆骨架物料排序扫描 ");
        intent.putExtra("url","");
        intent.setClass(activity, OneActivity.class);
        activity.startActivity(intent);
        break;
    case 3:
        onTakePhoto();
        intent.putExtra("title"," 前排靠背骨架物料排序扫描 ");
        intent.putExtra("url","");
        intent.setClass(activity, OneActivity.class);
        activity.startActivity(intent);
        break;
    case 4:
        onTakePhoto();
        intent.putExtra("title"," 前排线束物料排序扫描 ");
        intent.putExtra("url","");
        intent.setClass(activity, OneActivity.class);
```

```
                    activity.startActivity(intent);
                    break;
            case 5:
                    onTakePhoto();
                    intent.putExtra("title"," 后排 40% 靠背面套物料排序扫描 ");
                    intent.putExtra("url","");
                    intent.setClass(activity, OneActivity.class);
                    activity.startActivity(intent);
                    break;
            case 6:
                    onTakePhoto();
                    intent.putExtra("title"," 后排 60% 靠背面套物料排序扫描 ");
                    intent.putExtra("url","");
                    intent.setClass(activity, OneActivity.class);
                    activity.startActivity(intent);
                    break;
            case 7:
                    onTakePhoto();
                    intent.putExtra("title"," 后排坐垫面套物料排序扫描 ");
                    intent.putExtra("url","");
                    intent.setClass(activity, OneActivity.class);
                    activity.startActivity(intent);
                    break;
            case 8:
                    onTakePhoto();
                    intent.putExtra("title"," 插单物料排序扫描 ");
                    intent.putExtra("url","");
                    intent.setClass(activity, OneActivity.class);

                    activity.startActivity(intent);
                    break;
            }           }           });         }
```

第十五步：获取零件信息，填充列表具体代码如 CORE0224 所示。

代码 CORE0224　　将得到的零件信息填入列表
public class OneActivity extends AppCompatActivity { // 初始化控件

```java
private RelativeLayout title_bar;
private TextView tv_back, tv_main_title;
private LinearLayout ll_false,ll_lv;
private Button btn_close;
String url;
okhttp okhttp = new okhttp();  // 调用网络接口
private static final int REQUEST_CODE_SCAN = 0x0000;
private ListView lv_id;
// 定义填充列表的内容
List<String> list_id = new ArrayList<>();
List<String> list_carid = new ArrayList<>();
List<String> list_cardtype = new ArrayList<>();
List<String> list_mian = new ArrayList<>();
List<String> list_code = new ArrayList<>();
List<String> list_number = new ArrayList<>();
List<String> list_codenum = new ArrayList<>();
@Override
protected void onCreate(Bundle savedInstanceState) {
    super.onCreate(savedInstanceState);
    setContentView(R.layout.activity_one);
    initview();
}
//handler 线程操作
Handler handler = new Handler() {
    @Override
    public void handleMessage(Message msg) {
        switch (msg.what) {
            case 0:
                String list_string = (String) msg.obj;
                if (!list_string.equals("")) {
                    JsonData(list_string);
                }
                break;        }    }   };
// 自定义 JSON 解析方法
public void JsonData(String aa) {
    Gson gson = new Gson();  // 调用 Gson 类
    Root rt = gson.fromJson(aa, Root.class);// 得到 Gson 串并使用实例化类解析
    Log.d("879798----------", rt.getList_string().size() + "---");
```

```java
        // 循环添加数据在列表
        for (int i = 0; i < rt.getList_string().size(); i++) {
            List_string data = rt.getList_string().get(i);
            String tv_id = data.getTv_id();
            String tv_carnum = data.getTv_carnum();
            String tv_icartype = data.getTv_icartype();
            String tv_main = data.getTv_main();
            String tv_code = data.getTv_code();
            String tv_num = data.getTv_num();
            String tv_lookcode = data.getTv_lookcode();
            list_id.add(tv_id);
            list_carid.add(tv_carnum);
            list_cardtype.add(tv_icartype);
            list_mian.add(tv_main);
            list_code.add(tv_code);
            list_number.add(tv_num);
            list_codenum.add(tv_lookcode);
        }
        // 使用适配器进行刷新
        OneAdapter adapter = new OneAdapter(OneActivity.this,
            list_id, list_carid, list_cardtype, list_mian, list_code,
            list_number, list_codenum);
        lv_id.setAdapter(adapter);
    }
    private void initview() {
        // 获取控件内容
        final OneAdapter.ViewHolder holder;
        Intent intent = getIntent();
        // 得到上个 activity 传来的数据
        String title = intent.getStringExtra("title");
        String content = intent.getStringExtra("context");
        url = intent.getStringExtra("url");
        title_bar = (RelativeLayout) findViewById(R.id.title_bar);
        title_bar.setBackgroundColor(Color.parseColor("#30B4FF"));
        tv_main_title = (TextView) findViewById(R.id.tv_main_title);
        tv_main_title.setText(title);
        lv_id = (ListView) findViewById(R.id.lv_id);
```

```java
// 给缓存文件设置路径
File sdcache = this.getExternalCacheDir();
okhttp.initOkHttpClient(sdcache);
try {
    // 请求数据
    RequestBody body = new FormBody.Builder().
            add("JSON", "{flag:" + content + "}").build();
    okhttp.postAsynHttp(url, body, "list_string");
    handler.postDelayed(new Runnable() {
        @Override
        public void run() {
            // 得到数据并进行 handler 抛出处理
            String  str= Info.map.get("list_string");
            Message msg = new Message();
            msg.what = 0;
            msg.obj = str;
            handler.sendMessage(msg);
        }
    }, 500);
} catch (Exception e) {
}
tv_back = (TextView) findViewById(R.id.tv_back);
tv_back.setOnClickListener(new View.OnClickListener() {
    @Override
    public void onClick(View v) {
        OneActivity.this.finish();
    }
});
}}
```

第十六步：点击扫描进行零件号的读取，具体代码如 CORE0225 所示。

代码 CORE0225　将得到的零件信息填入列表

```java
public class OneActivity extends AppCompatActivity {
// 初始化控件及定义参数
private RelativeLayout title_bar;
private Button bt_scan;
String result;
String url,url1;
okhttp okhttp = new okhttp();
```

```java
private ListView lv_id;
// 初始化填充列表
List<String> list_id = new ArrayList<>();
List<String> list_carid = new ArrayList<>();
List<String> list_cardtype = new ArrayList<>();
List<String> list_mian = new ArrayList<>();
List<String> list_code = new ArrayList<>();
List<String> list_number = new ArrayList<>();
List<String> list_codenum = new ArrayList<>();
@Override
protected void onCreate(Bundle savedInstanceState) {
    super.onCreate(savedInstanceState);
    setRequestedOrientation(ActivityInfo.SCREEN_ORIENTATION_LANDSCAPE);
    setContentView(R.layout.activity_one);
    initview();
}
//handler 处理数据
Handler handler = new Handler() {
    @Override
    public void handleMessage(Message msg) {
        switch (msg.what) {
            case 1:
                String list_string1 = (String) msg.obj;
                if (!list_string1.equals("")) {
                    JsonData(list_string1);
                }
                break;
        }
    }
};
// 解析 JSON 数据
public void JsonData(String aa) {
    Gson gson = new Gson();                           // 调用 Gson 方法
    Root rt = gson.fromJson(aa, Root.class);          // 找到序列化类
    Log.d("879798----------", rt.getList_string().size() + "---");
    for (int i = 0; i < rt.getList_string().size(); i++) {  // 循环添加数据
        List_string data = rt.getList_string().get(i);
        String tv_id = data.getTv_id();
        String tv_carnum = data.getTv_carnum();
        String tv_icartype = data.getTv_icartype();
```

```
            String tv_main = data.getTv_main();
            String tv_code = data.getTv_code();
            String tv_num = data.getTv_num();
            String tv_lookcode = data.getTv_lookcode();
            list_id.add(tv_id);
            list_carid.add(tv_carnum);
            list_cardtype.add(tv_icartype);
            list_mian.add(tv_main);
            list_code.add(tv_code);
            list_number.add(tv_num);
            list_codenum.add(tv_lookcode);
        }
        // 适配器填充
        OneAdapter adapter = new OneAdapter(OneActivity.this, list_id, list_carid, list_cardtype, list_mian, list_code, list_number, list_codenum);
        lv_id.setAdapter(adapter);
    }
    private void initview() {
        // 获取控件及上个界面传来的信息
        Intent intent = getIntent();
        url = intent.getStringExtra("url");
        url1 = intent.getStringExtra("url1");
        tit tv_back = (TextView) findViewById(R.id.tv_back); // 返回上一个界面
        tv_back.setOnClickListener(new View.OnClickListener() {
            @Override
            public void onClick(View v) {
                startActivity(new Intent(OneActivity.this, activity.getClass()));
                OneActivity.this.finish();
            }      });
        bt_scan=(Button) findViewById(R.id.bt_scan); // 点击扫码（扫描取料零件号）
```

第十七步：再次跳转到扫描界面，进行再一次扫描，具体代码如 CORE0226 所示。

代码 CORE0226　再一次扫描零件

```
// 跳转到扫码界面
Intent intent = new Intent(OneActivity.this, CaptureActivity.class);
onTakePhoto();
activity.startActivityForResult(intent,REQUEST_CODE_SCAN);
```

```java
            try {
                Thread.sleep(3000);
            } catch (InterruptedException e) {
                e.printStackTrace();
            }               }               });       }
    // 动态获取照相机权限
    public void onTakePhoto()   {
    @Override
// 数据回调方法
protected void onActivityResult(int requestCode, int resultCode, Intent data) {
        super.onActivityResult(requestCode, resultCode, data);
if (requestCode == REQUEST_CODE_SCAN && resultCode == RESULT_OK) {
            if (data != null) {
                String content = data.getStringExtra("codedContent");
                System.out.print("++++++++content++++++++++++"+content);
                // 数据回调处理
                if (content!=null){
                    File sdcache = this.getExternalCacheDir();
                    // 获取服务器的返回数据
                    okhttp.initOkHttpClient(sdcache);
                    try {
                        RequestBody body = new FormBody.Builder().
                            add("JSON", "{flag:" + content + "}").build();
                        okhttp.postAsynHttp(url1, body, "list_string");
                        handler.postDelayed(new Runnable() {
                            @Override
                            //handler 线程抛出处理
                            public void run() {
                                String  str= Info.map.get("list_string");
                                Message msg = new Message();
                                msg.what = 1;
                                msg.obj = str;
                                handler.sendMessage(msg);
                            }                                   }, 500);
                    } catch (Exception e) {
                    }               }
                Log.d(" 解码结果 ", content + "content---------------------");
            }           }       }
```

运行程序,实现二维码/条形码扫描功能。

本模块介绍了此项目条形码扫描的实现,通过本模块的学习可以了解 Zxing 实现二维码扫描的使用方法,掌握通过 OkHttp 协议进行网络的数据传输。学习之后能够实现本项目的条形码扫描。

通常情况下,在权限管理系统中,权限仅仅在 App 安装时询问用户一次,用户同意了这些权限 App 才能被安装。在 Android6.0 开始,App 可以直接安装,App 在运行时一个一个询问用户授予权限,系统会弹出一个对话框让用户选择是否授权某个权限给 App,当 App 需要用户授予不恰当的权限的时候,用户可以拒绝,用户也可以在设置界面对每个 App 的权限进行管理。

技能扩展——Android 6.0 权限管理

1 权限说明

从 android 6.0(API23) 开始,系统的权限不会在安装的时候去请求,而是在程序运行时才会去请求。系统权限分为 2 种,分别为 normal 和 dangerous:

- Normal permission: 对于用户隐私没有危险的,可以在清单文件中直接添加的权限。
- Dangerous permission:app 需要访问用户的隐私信息等,即使在清单文件中已经添加授权,但是在程序运行时还是会请求用户授权。

2 检查权限 (check permission)

在 android6.0 及以上的 sdk 提供了一个检查权限的方法:ContextCompat.checkSelfPermission()。

```
int permission = ContextCompat.checkSelfPermission(MainActivity.this,
Manifest.permission.RECORD_AUDIO);
if (permission == PackageManager.PERMISSION_GRANTED) {
    // 表示已经授权
}
```

```
//PackageManager.PERMISSION_DENIED---> 表示权限被否认了
// 如果在 Activity 中申请权限可以的调用:
if (Build.VERSION.SDK_INT >= Build.VERSION_CODES.M) {
    int permission = checkSelfPermission(Manifest.permission.RECORD_AUDIO);
}
```

3 请求权限

在 Android 6.0 及以上,如果检查没有权限,需要主动请求权限。在请求权限时会弹出一个对话框提示用户是否授权。 请求权限的方法: requestPermissions()。在请求权限后会有一个回调方法 onRequestPermissionsResult()。

```
ActivityCompat.requestPermissions(MainActivity.this, new String[]{Manifest.permission.RECORD_AUDIO}, 1);
// 或者在 Activity 的方法调用
if (Build.VERSION.SDK_INT >= Build.VERSION_CODES.M) {
    requestPermissions(new String[]{Manifest.permission.RECORD_AUDIO},1);
}
/**
 * 参数 1: requestCode--> 是 requestPermissions() 方法传递过来的请求码。
 * 参数 2: permissions--> 是 requestPermissions() 方法传递过来的需要申请权限。
 * 参数 3: grantResults--> 是申请权限后,系统返回的结果,PackageManager.PERMISSION_GRANTED 表示授权成功, PackageManager.PERMISSION_DENIED 表示授权失败。
 * grantResults 和 permissions 是一一对应的
 */
@Override
public void onRequestPermissionsResult(int requestCode, @NonNull String[] permissions, @NonNull int[] grantResults) {
    super.onRequestPermissionsResult(requestCode, permissions, grantResults);
}
```

4 请求步骤

1. 将 targetSdkVersion 设置为 23,注意,如果将 targetSdkVersion 设置为 >=23,则必须按照 Android 谷歌的要求,动态申请权限,如果暂时不申请动态权限,则 targetSdkVersion 最大只能设置为 22。

2. 在 AndroidManifest.xml 中申请需要的权限,包括普通权限和需要申请的特殊权限。

3. 开始申请权限,此处分为 3 步,如下所示,权限请求流程如图 2.10 所示。

（1）检查是否有此权限 checkSelfPermission()，如果有且已经开启，则直接进行下一步操作。

（2）如果未开启，则判断是否需要向用户解释为何申请权限 shouldShowRequestPermissionRationale。

（3）如果需要（即返回 true），则可以弹出对话框提示用户申请权限原因，用户确认后申请权限 requestPermissions()，如果不需要（即返回 false），则直接申请权限 requestPermissions()。

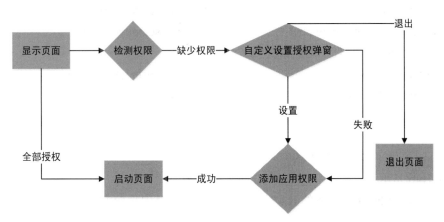

图 2.10　权限请求流程图

Accounts	账号	Cipher	密码
Directory	目录	Domain	域
Encryption	加密	Filter	过滤器
ISP	网络服务提供者	Password	口令
Permission	权限	Sender	发送者

一、选择题

1. Zxing 不支持以下哪种编码格式（　　）。
   A. UPC-A and UPC-E　　　　　　B. EAN-8 and EAN-13
   C. Code 40　　　　　　　　　　D. Code 39

2. 下列不属于 OKHttp 的优势是（　　）。
   A. 允许连接到同一个主机地址的所有请求，提高请求效率。
   B. 共享 Socket, 减少对服务器的请求次数 。

C. 缓存响应数据来减少重复的网络请求。

D. 增加对数据流量的消耗。

3. 为了更好地应对网络访问，OkHttp 不具有以下哪种功能（　　）。

A. Get 请求　　　　　　　　　　　　B. Post 请求

C. 基于 Http 的文件上传　　　　　　D. 不支持 session 的保持

4. onResponse 回调的参数是 response，一般情况下获得返回的字符串，通过（　　）获取。

A. response.body().string()　　　　　B. response.body().bytes()

C. response.body().byteStream()　　 D. response.body().int()

5. 下列是获取输入控件字符串的方法是（　　）。

A. getText()　　　　　　　　　　　　B. getEditString()

C. getSharedPreferences()　　　　　D. getActivity()

二、填空题

1. Zxing 提供了多种平台下的客户端包括：_____、_____ 和 Android。

2. OkHttp 是用来替代_____和 Apache HttpClient。

3. 将 Zxing android 目录下的 res 资源文件拷贝到项目中相应的位置，它会提示是否覆盖，选择_____。

4. Android 网络框架之 OKhttp 是一个_____的开源项目。

5. Okhttp 内部依赖 okio，同时导入_____。

三、上机题

1. 使用 Zxing 编写代码实现二维码功能。

2. 使用 Get 请求获取一个网页的内容。

模块三　扫描记录

通过实现获取历史记录，了解数据传输方式，学习请求与响应的相互关系，掌握历史记录的流程分配，具备独立完成获取扫描历史记录的能力。在任务实现过程中：

- 了解数据传输过程。
- 学习响应与请求的相互关系。
- 掌握历史记录的获取流程。
- 具备实现获取历史记录的能力。

项目一　物料排序手持端

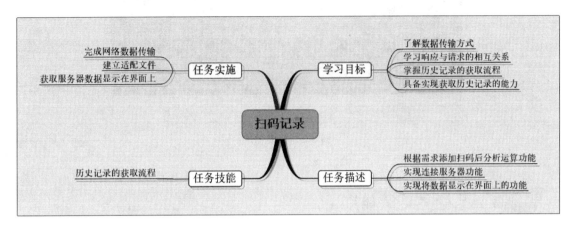

本模块将实现记录扫描结果的功能,用户在对零件扫描之后,服务器将扫描信息进行分析与运算后得到结果,并将扫描结果生成记录表,最后将其发送到客户端显示在界面上。

【功能描述】

本模块将实现此项目中的扫描记录模块。
- 使用 OkHttp 网络协议进行数据传输。
- 获取服务器的数据。

【基本框架】

基本框架如图 3.1 所示。

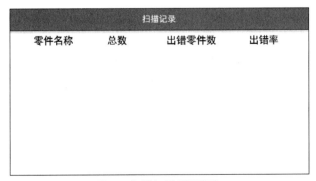

图 3.1 扫描记录界面框架

通过本模块的学习,将框架图转换成图 3.2 所示效果。

图 3.2　扫描记录

技能点一　历史记录获取流程

1　历史记录流程

Android 提供了两种 HTTP 交互的方式：HttpURLConnection 和 Apache HTTP Client，两者都支持 HTTPS、流的上传和下载、配置超时、IPv6 和连接池。虽然以上都可以满足各种 HTTP 请求的需求，但更高效的使用 HTTP 可以使应用运行更快、更节省流量。通过 OkHttp 库将扫描后的历史记录清晰明了地展现给用户，历史记录的时序图如 3.3 所示。

2　历史记录设计思路

OkHttp 总体设计如图 3.4 所示，Diapatcher 不断从 RequestQueue 中获取请求（Call），根据判断缓存是否存在调用 Cache 或 Network 这两类数据获取接口之一，最后从内存缓存或服务器取得请求的数据。该引擎有同步和异步请求，同步请求通过 Call.execute() 直接返回当前的 Response，而异步请求会把当前的请求 Call.enqueue 添加（AsyncCall）到请求队列中，并通过回调（Callback）的方式来获取最后结果。

根据历史记录总体设计后，详细请求流程如图 3.5 所示。

项目一 物料排序手持端

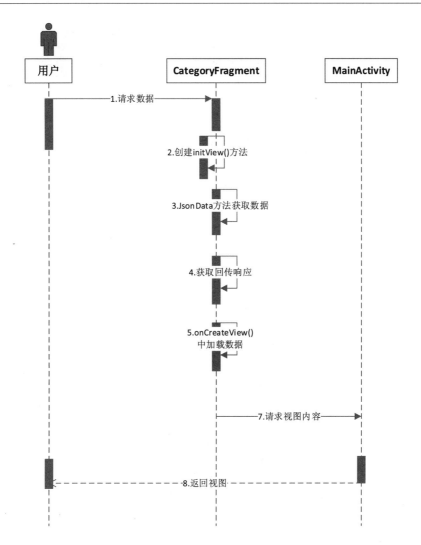

图 3.3 历史记录时序图

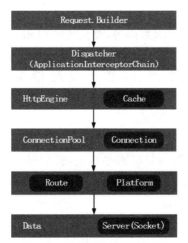

图 3.4 历史记录总体设计图

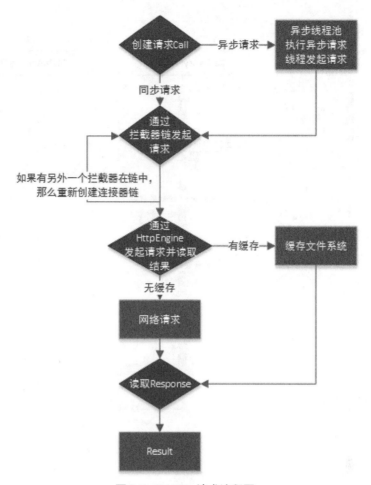

图 3.5 OkHttp 请求流程图

3 历史记录请求机制

历史记录由 OkHttp 提供,它支持各种版本的协议并且具有下面两种特性。

1. 多路复用机制

HttpEngine(Http 引擎)每次请求数据时,都会先调用 nextConnection(),如果返回一个连接对象,则就调用 sendRequest() 发送一个请求。如果 nextConnection() 返回为 null,就会调用 createNextConnection() 创建一个连接,然后发送请求。如果第一次调用 readNetworkResponse(),可能会返回一个 null,结束当前的调用,回到 nextConnection() 进行再一次连接。多路复用机制时序图如图 3.6 所示。

2. 重连机制

通过一个 while 的循环,判断当前状态是否连接,如果没有连接,调用 getResquese() 发送请求,紧接着 HttpEngine 调用 recover(),进行重连操作,直到当前连接状态为已连接,结束循环。重连机制时序图如图 3.7 所示。

项目一 物料排序手持端

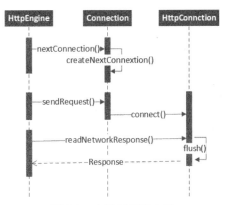

图 3.6 多路复用时序图

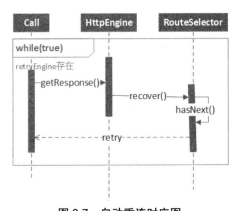

图 3.7 自动重连时序图

拓展：通过以上内容的学习，大家已经知道历史记录获取的流程、设计思路以及请求机制。在设计思路部分涉及到缓存，我们知道在正常情况下进入 app 首页后，图片加载完成，接着退出 app；然后断开网络，再进入 app 首页，界面显示空白，这是为什么呢？想知道具体的原因吗，扫描右侧二维码，了解 Android 进阶之缓存机制与实现。

通过以上内容的学习，在已有项目的基础上完成扫描记录模块的开发。具体实现步骤如下所示。

第一步：通过网络 OkHttp 协议，进行网络的数据传输。具体代码如 CORE0301 所示。

```
代码 CORE0301    OkHttp 实体类
public class okhttp {
public OkHttpClient mOkHttpClient;
public String str;
```

```java
// 构建 initOkHttpClient
public void initOkHttpClient(File sdcache) {
    int cacheSize = 10 * 1024 * 1024;
    OkHttpClient.Builder builder = new OkHttpClient.Builder()
            .connectTimeout(15, TimeUnit.SECONDS) // 设置连接的超时时间
            .writeTimeout(20, TimeUnit.SECONDS)   // 设置响应的超时时间
            .readTimeout(20, TimeUnit.SECONDS)    // 请求的超时时间
            .cache(new Cache(sdcache.getAbsoluteFile(), cacheSize));
    mOkHttpClient = builder.build();
}
/**
 * post 请求
 * @param url      请求 url
 * @param callback 请求回调
 * @param formBody 请求参数
 */
public void postAsynHttp(final String url, final RequestBody formBody, final String key) {
    Log.d("cookie.url", Info.cookie + "-----str------");
    // 请求数据前的请求体
    Request request = new Request.Builder()
            .url(url)
            .post(formBody)
            .addHeader("cookie", Info.cookie)
            .build();
    Call call = mOkHttpClient.newCall(request);   // 数据回调
    call.enqueue(new Callback() {
        @Override
        public void onFailure(Call call, IOException e) {
            Log.d("e", e + "-------");
        }
        @Override
        public void onResponse(Call call, Response response) throws IOException {
            if (response.isSuccessful()) {
                str = response.body().string(); // 把传输数据转化为 string 类型
                if (!TextUtils.isEmpty(str)) {
                    if (url == Url.login) {
                        List<String> cookieList = response.headers("Set-Cookie");
```

```
                        // 接收并放入 List 列表中
                        String session = cookieList.get(0);
                        // 截取数据
                        session = session.substring(0, session.indexOf(";"));
                        // 将得到的数据进行截取
                        Info.cookie = session;
                    }
                    Info.map.put(key, str);        // 数据填充
                    Log.d("str", str + "-----str------");}
            } else {
                Log.d("response", response.isSuccessful() + "-------------");
                Boolean flog = response.isSuccessful();
                Info.map.put(key, flog + "");
    } }    }); }
```

第二步：编写扫描记录的适配器布局文件，如图 3.8 所示。

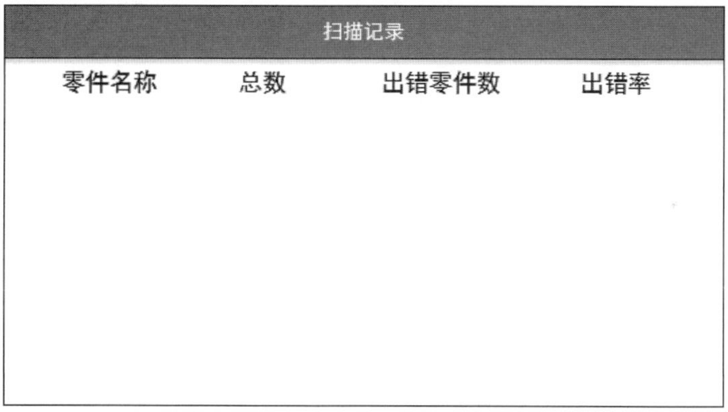

扫描记录			
零件名称	总数	出错零件数	出错率

图 3.8 扫描记录界面框架图

第三步：建立适配器文件。具体代码如 CORE0302 所示。

代码 CORE0302 适配器的编写

```
public class CategoryAdapter extends BaseAdapter {
// 数据集合
List<String> list_codename;
List<String> list_numbers;
List<String> list_fcodenum;
List<String> list_flases;
Context context;
// 构造器
```

```java
public CategoryAdapter(List<String> list_codename, List<String> list_numbers,
    List<String> list_fcodenum, List<String> list_flases, Context context) {
    this.list_codename = list_codename;
    this.list_numbers = list_numbers;
    this.list_fcodenum = list_fcodenum;
    this.list_flases = list_flases;
    this.context = context;
}
// 返回已定义数据源总数量
@Override
public int getCount() {
    return list_codename.size();
}
// 告诉适配器取得目前容器中的数据对象
@Override
public Object getItem(int position) {
    return list_codename.get(position);
}
// 告诉适配器取得目前容器中的数据 ID
@Override
public long getItemId(int position) {
    return position;
}
// 实现 getView 方法
@Override
public View getView(int position, View convertView, ViewGroup parent) {
    // 反射行布局
    convertView= LayoutInflater.from(context).inflate(
    R.layout.category_adapter,null);
    // 获取各个控件
    TextView tv_codename=(TextView) convertView.findViewById(
    R.id.tv_codename);
    TextView tv_numbers=(TextView) convertView.findViewById(
    R.id.tv_numbers);
    TextView tv_fcodenum=(TextView) convertView.findViewById
    (R.id.tv_fcodenum);
    TextView tv_flases=(TextView) convertView.findViewById(R.id.tv_flases);
    // 给控件赋值
```

```
            tv_codename.setText(list_codename.get(position));
            tv_numbers.setText(list_numbers.get(position));
            tv_fcodenum.setText(list_fcodenum.get(position));
            tv_flases.setText(list_flases.get(position));
            return convertView;
        }   }
```

第四步：获取服务器数据，显示在界面上，实现如图 3.9 所示效果。具体代码如 CORE0303 所示。

代码 CORE0303　获取服务器数据

```
public class CategoryFragment extends Fragment {
    // 初始化控件
    private View mView;
    private RelativeLayout title_bar;
    private TextView tv_back,tv_main_title;
    private ListView lv_record;
    // 初始化 okhttp 类
    okhttp okhttp=new okhttp();
    // 初始化 list 列表
    List<String> list_codename=new ArrayList<>();
    List<String> list_numbers=new ArrayList<>();
    List<String> list_fcodenum=new ArrayList<>();
    List<String> list_flases=new ArrayList<>();
    @Nullable
    @Override
    public View onCreateView(LayoutInflater inflater,
        @Nullable ViewGroup container, @Nullable Bundle savedInstanceState) {
        mView=inflater.inflate(R.layout.fragment_category,null);
        initview();
        return mView;
    }
    //handler 处理数据
    Handler handler=new Handler(){
        @Override
        public void handleMessage(Message msg) {
            switch (msg.what){
                case 0:
```

```java
                    String recode=(String) msg.obj;
                    if (recode!=null){
                    JsonData(recode);
                    }
                    break;
        }  }      };
private void initview() {          // 获取布局中的控件
    title_bar=(RelativeLayout) mView.findViewById(R.id.title_bar);
    title_bar.setBackgroundColor(Color.parseColor("#30B4FF"));
    tv_main_title=(TextView)mView.findViewById(R.id.tv_main_title);
    tv_main_title.setText(" 通讯录 ");
    tv_back=(TextView)mView.findViewById(R.id.tv_back);
    tv_back.setVisibility(View.GONE);
    lv_record=(ListView) mView.findViewById(R.id.lv_record);
    // 缓存路径
    File sdcache = getActivity().getExternalCacheDir();
    okhttp.initOkHttpClient(sdcache);
    try {// 构造请求体
            RequestBody body = new FormBody.Builder().
                    add("JSON", "{flag:code}").build();
    // 获取请求内容
        okhttp.postAsynHttp("http://192.168.2.105:8080/SSMDemo/users/pro15 ",
        body, "Strings");
        handler.postDelayed(new Runnable() {// 将获取的内容进行 handler 抛出处理
                @Override
                public void run() {
                    String  str= Info.map.get("Strings");
                    Message msg = new Message();
                    msg.what = 0;
                    msg.obj = str;
                    handler.sendMessage(msg);
                }
            }, 500);
    } catch (Exception e) {
    }  }
//JSON 串的解析方法
public void JsonData(String aa) {
```

```
Gson gson = new Gson();
JsonRootBean rt = gson.fromJson(aa, JsonRootBean.class);
Log.d("879798----------", rt.getStrings().size() + "---");
for (int i = 0; i < rt.getStrings().size(); i++) {
        Strings data= rt.getStrings().get(i);
        String tv_codename = data.getCodename();
        String tv_numbers = data.getNumbers();
        String tv_fcodenum = data.getFcodenum();
        String tv_flases = data.getFalses();
        list_codename.add(tv_codename);
        list_numbers.add(tv_numbers);
        list_fcodenum.add(tv_fcodenum);
        list_flases.add(tv_flases);
    }
// 调用自定义适配器
CategoryAdapter adapter = new CategoryAdapter(list_codename,
    list_numbers,List_fcodenum,list_flases,getActivity());
    lv_record.setAdapter(adapter);
}   }
```

图 3.9 扫描记录界面效果

本模块介绍了此项目扫描记录的实现,通过本模块的学习可以熟练的运用 OkHttp 协议进行网络传输,并可以从中理解到请求与响应的过程。学习之后可以与服务器之间进行简单的信息传递。

技能扩展——Volley

1　简介

在 2013 年 Google I/O 大会上推出了一个新的网络通信框架 Volley。Volley 既可以访问网络获取数据，也可以加载图片，并且在性能方面也进行了大幅度的调整，其非常适合进行数据量不大，但通信频繁的网络操作。而对于大数据量的网络操作，比如说下载文件等，Volley 的表现就会非常糟糕。在使用 Volley 前请下载 Volley 库放在 libs 目录下并加入 add 到工程中。

2　Volley 的功能

- Json，图像等异步下载。
- 网络请求的排序（scheduling）。
- 网络请求的优先级处理。
- 缓存。
- 多级别取消请求。
- 和 Activity 的生命周期联动（Activity 结束的同时取消所有网络请求）。

3　Volley 的优缺点

如表 3.1 所示是 Volley 的优缺点。

表 3.1　Volley 优缺点

优点	缺点
非常适合进行数据量不大，但通讯频繁的操作	使用的是 HttpClient、HTTPURLConnection
可直接在主线程调用服务端并处理返回结果	6.0 不支持 HttpClient，如果想支持需要添加 oprg.apache.http.legacy.jar
可以取消请求，容易扩展，面向接口编程	对大文件下载 Volley 的表现非常糟糕
网络请求线程 Network Dispatcher 默认开启了 4 个，可以优化	只支持 Http 请求
通过使用标准的 HTTP 缓存机制保持磁盘和内存响应的一致	图片加载性能一般

4　使用 Volley 加载数据

1. StringResponse 的用法

首先通过调用如下方法获取到一个 RequestQueue 对象：

```
RequestQueue mQueue = Volley.newRequestQueue(context);
```

发送一条 HTTP 请求，创建一个 StringResponse 对象，如下所示：

```
StringRequest stringRequest = new StringRequest("http://www.baidu.com",
                new Response.Listener<String>() {
                    @Override
                    public void onResponse(String response) {
                        Log.d("TAG", response);
                    }
                }, new Response.ErrorListener() {
                    @Override
                    public void onErrorResponse(VolleyError error) {
                        Log.e("TAG", error.getMessage(), error);
                    }        });
```

可以看到，这里 new 出了一个 StringRequest 对象， StringRequest 的构造函数需要传入三个参数，第一个参数就是目标服务器的 URL 地址，第二个参数是服务器响应成功的回调，第三个参数是服务器响应失败的回调。其中，目标服务器地址是百度的首页，在响应成功的回调里打印出服务器返回的内容，在响应失败的回调里打印失败的详细信息。

最后，将这个 StringRequest 对象添加到 RequestQueue 中，如下所示：

```
mQueue.add(stringRequest);
```

运行程序，并发出 HTTP 请求，就会看到 LogCat 中打印出如图 3.10 所示的数据。

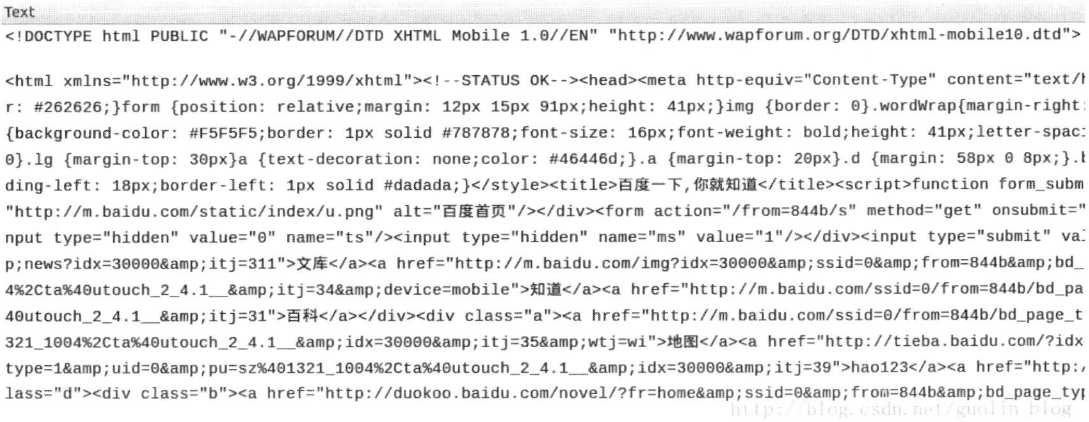

```
Text
<!DOCTYPE html PUBLIC "-//WAPFORUM//DTD XHTML Mobile 1.0//EN" "http://www.wapforum.org/DTD/xhtml-mobile10.dtd">

<html xmlns="http://www.w3.org/1999/xhtml"><!--STATUS OK--><head><meta http-equiv="Content-Type" content="text/
r: #262626;}form {position: relative;margin: 12px 15px 91px;height: 41px;}img {border: 0}.wordWrap{margin-right
{background-color: #F5F5F5;border: 1px solid #787878;font-size: 16px;font-weight: bold;height: 41px;letter-spac
0}.lg {margin-top: 30px}a {text-decoration: none;color: #46446d;}.a {margin-top: 20px}.d {margin: 58px 0 8px;}.
ding-left: 18px;border-left: 1px solid #dadada;}</style><title>百度一下,你就知道</title><script>function form_subm
"http://m.baidu.com/static/index/u.png" alt="百度首页"/></div><form action="/from=844b/s" method="get" onsubmit="
nput type="hidden" value="0" name="ts"/><input type="hidden" name="ms" value="1"/></div><input type="submit" va
p;news?idx=30000&amp;itj=311">文库</a><a href="http://m.baidu.com/img?idx=30000&amp;ssid=0&amp;from=844b&amp;bd_
4%2Cta%40utouch_2_4.1__&amp;itj=34&amp;device=mobile">知道</a><a href="http://m.baidu.com/ssid=0/from=844b/bd_pa
40utouch_2_4.1__&amp;itj=31">百科</a></div><div class="a"><a href="http://m.baidu.com/ssid=0/from=844b/bd_page_t
321_1004%2Cta%40utouch_2_4.1__&amp;idx=30000&amp;itj=35&amp;wtj=wi">地图</a><a href="http://tieba.baidu.com/?idx
type=1&amp;uid=0&amp;pu=sz%401321_1004%2Cta%40utouch_2_4.1__&amp;idx=30000&amp;itj=39">hao123</a><a href="http:
lass="d"><div class="b"><a href="http://duokoo.baidu.com/novel/?fr=home&amp;ssid=0&amp;from=844b&amp;bd_page_ty
```


图 3.10　打印出的数据

2. JsonRequest 的用法

JsonRequest 也继承了 Request 类，由于 JsonRequest 是一个抽象类，因此无法进行实例化操作，那么只能从它的子类去开始。JsonRequest 有两个直接的子类，JsonObjectRequest 和 JsonArrayRequest，一个用来请求 JSON 数据，一个是用来请求 JSON 数组。

首先 new 出一个 JsonObjectRequest 对象，示例如下所示：

```
JsonObjectRequest jsonObjectRequest = new JsonObjectRequest("http://m.weather.com.cn/data/101010100.html", null,
    new Response.Listener<JSONObject>() {
        @Override
        public void onResponse(JSONObject response) {
            Log.d("TAG", response.toString());
        }
    }, new Response.ErrorListener() {
        @Override
        public void onErrorResponse(VolleyError error) {
            Log.e("TAG", error.getMessage(), error);
        }
    });
```

可以看到，一个 URL 地址 http://m.weather.com.cn/data/101010100.html，这是中国天气网提供的一个可以查询天气信息的接口，数据的响应就是以 JSON 格式返回的，然后在 onResponse() 方法中将返回的数据打印出来。

最后再将这个 JsonObjectRequest 对象添加到 RequestQueue 里，如下所示：

```
mQueue.add(jsonObjectRequest);
```

运行一下程序，发出 HTTP 请求，在 LogCat 中会打印出如图 3.11 所示的数据。

```
Text
{"weatherinfo":{"weather6":"多云","weather5":"多云","weather4":"晴转多云","index_d":"天气寒冷,
9","img1":"0","index":"寒冷","tempF1":"46.4下~26.6下","img_title10":"多云","img_title11":"多云
14年3月4日","city_en":"beijing","index48_d":"天气冷, 建议着棉服、羽绒服、皮夹克加羊毛衫等冬季服装。年
"北风","st5":"7","st6":"-1","st3":"8","date":"","st4":"0","st1":"7","st2":"-3","temp1":"8℃~
4下~30.2下","tempF5":"50下~33.8下","index_ls":"基本适宜","tempF2":"46.4下~26.6下","tempF3":"44.
"fl4":"小于3级","temp6":"10℃~2℃","fl3":"小于3级","temp5":"10℃~1℃","fl2":"小于3级","temp4":"8
title5":"晴","img_title4":"晴","fchh":"11","img_title9":"多云","img10":"99","img_title8":"多
"wind2":"微风","weather3":"晴","wind5":"微风","img_title3":"晴","index_uv":"中等","wind4":"微
                                                  http://blog.csdn.net/guolin_blog
```

图 3.11 打印出的数据

由此可以看出，服务器返回的数据是以 JSON 格式显示的，并且 onResponse() 方法中携带的参数也正是一个 JSONObject 对象，最后只需要从 JSONObject 对象中找出所需要的数据即可。

3. 添加请求 add(Request) 及其工作流程

添加请求 add(Request) 流程如图 3.12 所示。

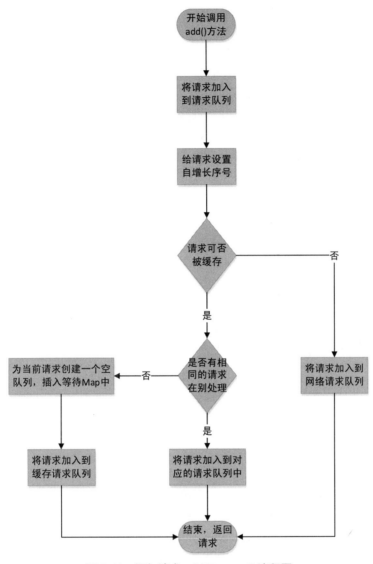

图 3.12　添加请求 add(Request) 流程图

具体示例如下所示。

```
public <T> Request<T> add(Request<T> request) {
/*
 * 将请求加入到当前请求队列当中，毋庸置疑的我们需要将所有的请
 * 求集合在一个队列中，方便最后做统一操作，例如：取消单个请求或
 * 者取消具有相同标记的请求
 */
```

```java
        request.setRequestQueue(this);
        synchronized (mCurrentRequests) {
            mCurrentRequests.add(request);
        }
        // 给请求设置顺序。
        request.setSequence(getSequenceNumber());
        request.addMarker("add-to-queue");
        // 如果请求是不能够被缓存的,直接将该请求加入网络队列中。
        if (!request.shouldCache()) {
            mNetworkQueue.add(request);
            return request;
        }
        /*
         * 如果有相同的请求正在被处理,就将请求加入对应请求的等待队列中
         * 去。等到相同的正在执行的请求处理完毕的时候会调用 finish() 方
         * 法,然后将这些等待队列中的请求全部加入缓存队列中去,让缓存线
         * 程来处理
         */
        synchronized (mWaitingRequests) {
            String cacheKey = request.getCacheKey();
            if (mWaitingRequests.containsKey(cacheKey)) {
                // 有相同请求在处理,加入等待队列。
                Queue<Request<?>> stagedRequests = mWaitingRequests.get(cacheKey);
                if (stagedRequests == null) {
                    stagedRequests = new LinkedList<>();
                }
                stagedRequests.add(request);
                mWaitingRequests.put(cacheKey, stagedRequests);
                if (VolleyLog.DEBUG) {
                    VolleyLog.v("Request for cacheKey=%s
                    is in flight, putting on hold.", cacheKey);}
            } else {
                /*
                 * 向 mWaitingRequests 中插入一个当前请求的空队列,
                 * 表明当前请求正在被处理
                 */
                mWaitingRequests.put(cacheKey, null);
                mCacheQueue.add(request);
```

```
            }
                return request;
        }   }
```

Cache	贮藏	Route	航线
Platform	平台	Dispatcher	调度
Response	回答	Result	结果
Flash	闪光	Recover	恢复
Position	位置	Record	记录

一、选择题

1. 通过 OkHttp 的学习判断以下说法正确的是（   ）。

   A. Httpengine.java 这个类主要是管理 HTTP 和 SPDY 连接时减少网络延迟

   B. okhttpclient.java 这个类主要是做平台适应性，针对 Android2.3 到 5.0 后的网络请求的适配支持

   C. 如果服务器配置了多个 IP 地址，当第一个 IP 连接失败的时候，OkHttp 会自动尝试下一个 IP

   D. OkHttp 支持各种版本的协议具有一种特性

2. 下列选项中不属于 OkHttp 中类的是（   ）。

   A. Platfrom.java   B. Connection.java

   C. Dispatcher.java   D. Callback.java

3. 根据任务实施的内容在下列选项中选出设置连接的超时时间的选项（   ）。

   A. connectTimeout(15, TimeUnit.SECONDS)

   B. writeTimeout(20, TimeUnit.SECONDS)

   C. readTimeout(20, TimeUnit.SECONDS)

   D. cache(new Cache(sdcache.getAbsoluteFile(), cacheSize))

4. 以下选项是 post 请求中请求参数的注解是（   ）。

   A. @param url   B. @param callback

   C. @param formBody   D. @Override

5. 根据任务实施判断下列选项，其中描述正确的是（   ）。

   A. // 告诉适配器取得目前容器中的数据 ID

```
    @Override
    public int getCount() {
     return list_codename.size();
    }
B. // 告诉适配器取得目前容器中的数据对象
    @Override
    public Object getItem(int position) {
    return list_codename.get(position);
    }
C. // 实现 getView 方法
    @Override
    public long getItemId(int position) {
    return position;
    }
D. // 返回已定义数据源总数量
    @Override
    public View getView(int position, View convertView, ViewGroup parent) {…}
```

二、填空题

1. 更高效的使用_____可以使应用运行更快、更节省流量。

2. OkHttp 的总体设计图,通过 Diapatcher 不断从 RequestQueue 中取出请求(Call),根据是否已缓存调用_____或_____这两类数据获取接口之一。

3. HttpEngine(Http 引擎)每次请求数据时,都会先调用_____,如果返回一个连接对象,则就调用_____发送一个请求。

4. 在多路复用机制中如果第一次调用 readNetworkResponse(),可能会返回一个_____,那么就结束当前的调用,回到_____进行再一次连接。

5. 重连机制是通过一个_____的循环,判断条件是当前状态是否连接,如果没有连接,就会调用_____发送请求,紧接着 HttpEngine 调用_____,进行重连操作,直到当前连接状态为已连接,结束循环。

三、上机题

1. 通过 OkHttp 实体类获取网络的天气数据。
2. 通过网络协议可以与服务器进行模拟聊天的过程。

项目二　新闻天下

模块一　新闻阅读

通过新闻阅读功能的实现，了解自定义控件的写法及属性，学习导航栏、导航栏切换、水平显示等自定义控件的用法，掌握滑动菜单控件的使用，具有自主编写自定义控件的能力。在任务实现过程中：
- 了解 Toolbar 控件的使用。
- 学习侧滑菜单实现方法。
- 熟悉下拉刷新实现功能。
- 掌握编写自定义控件的技能。

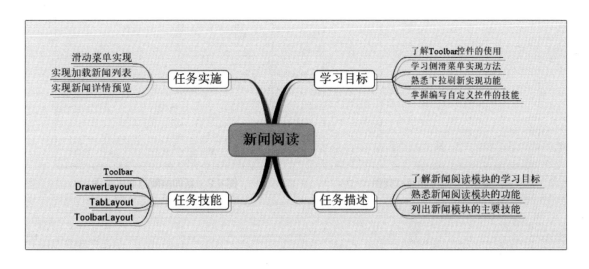

互联网极高的普及率直接影响着人们的生活,闲暇之余浏览时事新闻成为人们生活中不可缺少的部分。新闻天下充分从用户的角度出发进行设计,新闻浏览是新闻天下的核心模块,本模块主要是给用户提供当前最新的各种新闻资讯。本模块主要分为四个专区:头条、NBA、汽车、笑话,通过 JSON 数据来获取各类新闻资讯。用户可在感兴趣的模块点击列表项浏览新闻详情。

【功能描述】

本模块将实现新闻天下项目中的新闻阅读模块。
- 使用 Navigation View 实现侧滑菜单。
- 使用 Fragment 实现导航切换。
- 使用 JSON 数据获取新闻资讯。
- 实现下拉加载新闻。
- 实现新闻详情预览。

【基本框架】

基本框架如图 4.1 至 4.4 所示。

图 4.1 侧拉界面框架图

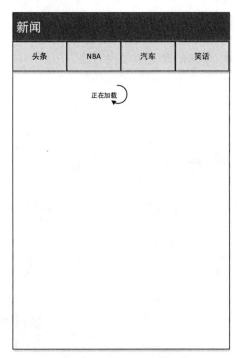

图 4.2 新闻刷新界面框架图

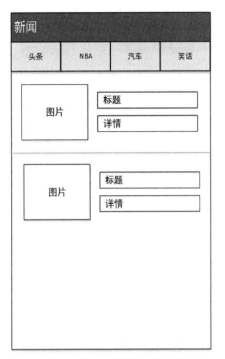
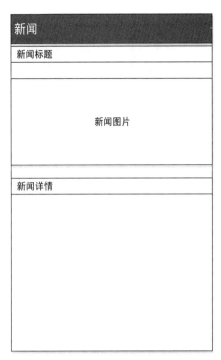

图 4.3　新闻列表界面框架图　　　　图 4.4　新闻详情界面框架图

通过本模块的学习,将以上的框架图转换成图 4.5 至图 4.8 所示效果。

图 4.5　侧拉界面效果图　　　　图 4.6　新闻刷新界面效果图

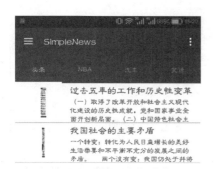

图 4.7 新闻列表界面效果图　　　　图 4.8 新闻详情界面效果图

谷歌在 2014 年推出一套全新的界面设计语言——Material Design，Material Design 有着完整的交互和视觉设计风格，具备视觉效果、运动元素、互动效果等特性。随着 Android5.0 的发布，Material Design 势必成为未来 Android 设计发展趋势。在 2015 年 Google I/O 大会上谷歌推出了 Design Support 库，将 Material Design 中的一些控件和效果进行了封装。其中常见的控件如表 4.1 所示。

表 4.1　Material Design 常见控件

控件	含义
Toolbar	继承了 ActionBar 的所有功能，使用更加灵活
AppBarLayout	继承 LinerLayout 实现的一个 ViewGroup 容器组件，支持手势滑动操作
DrawerLayout	为滑动菜单，通过滑动的方式将菜单显示出来
NavigationView	在滑动菜单界面定制任意的布局
SwipeRefreshLayout	用于实现下拉刷新功能的核心类
CollapsingToolbarLayout	实现一个可叠式标题栏的效果
RecyclerView	用于显示大量数据的控件，类似于 ListView、GirdView
CoordinatorLayout	从另一层面去控制子 view 之间触摸事件的布局
TabLayout	实现了固定的选项卡，也实现了可滚动的选项卡

本项目在编写的过程中应用到了 Material Design 中的一些控件,以下是对部分控件的介绍。

技能点一　Toolbar

1　Toolbar 简介

Toolbar 不仅继承了 ActionBar 的所有功能,而且使用更加灵活,可以自由的添加子控件,不像 ActionBar 那么固定,还可以配合其他控件完成 Material Design 的效果,显示的各种效果通过相应的属性方法即可实现。

2　Toolbar 的属性及效果图

1. Toolbar 的常用属性

Toolbar 常用的属性如表 4.2 所示。

表 4.2　Toolbar 常用属性

属性	含义
colorPrimaryDark	状态栏的颜色(可用来实现沉浸效果)
colorPrimary	Toolbar 的背景颜色
colorAccent	EditText 在输入时,RadioButton 选中时的颜色
android:textColorPrimary	Toolbar 中文字的颜色
app:title	Toolbar 中的 App Title
app:subtitle	Toobar 中的小标题
app:navigationIcon	导航图标

2.Toolbar 元素

Toolbar 包含以下元素:
- 导航按钮。
- 项目的 logo。
- 标题和子标题。
- 若干个自定义 View。
- 动作菜单。

通过学习以上属性,实现 Toolbar 功能,运行效果如图 4.9 所示。

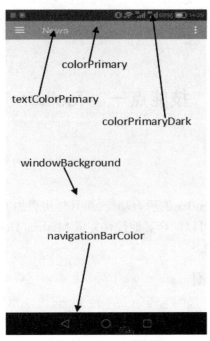

图 4.9　Toolbar 效果图

3. Toolbar 与 ActionBar

从 Android3.0(API11) 开始，所有使用默认主题的 Activity 都自带一个 ActionBar，但是随着 Android 版本的更新，ActionBar 的特性不断增加，从而导致了在不同 Android 版本的设备上，ActionBar 的显示不一致。

从 Android5.0(API 21) 开始，引进了 Toolbar，包含了 ActionBar 最新添加的多数特性，同时添加到了支持库中，使得在低版本设备上也可以使用 Toolbar。

Toolbar 与 ActionBar 的区别：
- ToolBar 是一个 View，与其他 View 一样包含在布局中。
- 与 View 一样，Toolbar 很容易用来放置、实现动画控制。
- 一个 Activity 中可以有多个 Toolbar。

4. Toolbar 的实现步骤

打开 res/values/styles.xml 文件，在 android：theme 属性更改一个不带 ActionBar 的主题。具体代码如 CORE0401 所示。

代码 CORE0401　指定 NoActionBar 的主题

```
<style name="AppTheme" parent="Theme.AppCompat.Light.NoActionBar">
```

在布局中定义一个 Toolbar 控件。具体代码如 CORE0402 所示。

代码 CORE0402　定义 Toolbar 布局

```
<android.support.v7.widget.Toolbar
android:layout_width="match_parent"
```

```
android:layout_height="40dp"
android:background="@color/color6"
app:popupTheme="@style/ThemeOverlay.AppCompat.Light" >
</android.support.v7.widget.Toolbar>
```

使用 Toolbar 来替代 ActionBar。具体代码如 CORE0403 所示。

代码 CORE0403　替代 ActionBar

```
Toolbar toolbar=(Toolbar)findViewById(R.id.toolbar);    // 调用 Toolbar
setSupportActionBar(toolbar);                           //ToolBar 控件替代 ActionBar 控件
```

技能点二　DrawerLayout

1 DrawerLayout 简介

DrawerLayout 是 V4 包下提供的一种左/右滑抽屉布局效果，是 Material Design 中最常见的效果之一。DrawerLayout 分为侧边菜单（NavigationView）和主内容区两部分，侧边菜单 NavigationView 可以根据手势展开与隐藏，主内容区的内容可以随着菜单的点击而变化。

2 NavigationView 简介

NavigationView 是一个导航 View。一般用它和 DrawerLayout 实现抽屉式导航设计，可以在滑动菜单界面定制任意的布局，这个菜单整体上分为两部分，上部分为 HeaderLayout，用于显示头布局，下部分为点击项 menu，用于建立 MenuItem 选项的菜单。

3 NavigationView 的属性及效果图

NavigationView 常用到的属性如表 4.3 所示。

表 4.3　NavigationView 常用属性

属性	含义
android:fitsSystemWindows	设置状态栏透明化与否
android:layout_gravity	设置抽屉，也就是 NavigationView 从左边或是右边打开
app:menu	设置菜单内容的 xml 布局
app:headerLayout	设置 NavigationView 的 HeaderView 的 xml 布局文件

通过以上内容的学习，实现侧滑功能，运行效果如图 4.10 所示。

图 4.10 侧滑效果图

4　NavigationView 的实现步骤

在 budil.gradle 中需要引用 DesignSupport 库。具体代码如 CORE0404 所示。

代码 CORE0404　　配置 Design Support 库
compile'com.android.support:design:25.3.1' compile'de.hdodenhof:circleimageview:2.1.0'　// 实现图片原型化的功能

在 res 下创建一个 menu 包。创建 menu 来显示具体的菜单项。具体代码如 CORE0405 所示。

代码 CORE0405　　Menu.xml

`<menu xmlns:android="http://schemas.android.com/apk/res/android">`
`<!-- group 表示一个组、singl 表示组中的所有菜单项只能单选 -->`
`<group android:checkableBehavior="single">`
　　`<item`
　　　　`android:id="@+id/navigation_item_news"`
　　　　`android:icon="@drawable/news"`
　　　　`android:title=" 新闻 " />`
　　`</group> </menu>`

添加 header 布局,放置头像、用户名。具体代码如 CORE0406 所示。

代码 CORE0406　　HeaderLayout 布局

```xml
<LinearLayout
    xmlns:app="http://schemas.android.com/apk/res-auto"
    android:layout_width="match_parent"
    android:layout_height="192dp"
    android:background="?attr/colorPrimaryDark"
    android:gravity="center"
    android:orientation="vertical"
    android:theme="@style/ThemeOverlay.AppCompat.Dark">
    <!--自定义圆形头像控件 -->
    <de.hdodenhof.circleimageview.CircleImageView
        android:id="@+id/profile_image"
        android:layout_width="72dp"
        android:layout_height="72dp"
        android:src="@drawable/protrait"
        app:border_color="@color/primary_light"
        app:border_width="2dp" />
</LinearLayout>
```

DrawerLayout 是一个布局控件,通过 DrawerLayout 的规定布局方式完成布局,即可具有侧滑的效果。具体代码如 CORE0407 所示。

代码 CORE0407　　侧滑布局

```xml
<android.support.v4.widget.DrawerLayout
    xmlns:app="http://schemas.android.com/apk/res-auto"
    android:id="@+id/drawer_layout"
    android:layout_width="match_parent"
    android:layout_height="match_parent" >
    <android.support.design.widget.NavigationView
        android:id="@+id/nav_view"
        android:layout_width="match_parent"
        android:layout_height="match_parent"
        android:layout_gravity="start"
        app:headerLayout="@layout/nav_header"
        app:menu="@menu/nav_menu" />
</android.support.v4.widget.DrawerLayout>
```

实现滑屏功能。具体代码如 CORE0408 所示。

代码 CORE0408　实现滑动菜单功能

```
protected void onCreate(Bundle savedInstanceState) {
// 调用 findViewById() 方法得到了 Drawer Layout 的实例
mDrawerLayout=(DrawerLayout) findViewById(R.id.drawer_layout);
mNavigationView =(NavigationView)findViewById(R.id.nav_view);
// 在左上角创建一个旋转按钮
mDrawerToggle = new ActionBarDrawerToggle(this, mDrawerLayout,
toolbar, R.string.drawer_open, R.string.drawer_close);
mDrawerToggle.syncState();
mDrawerLayout.setDrawerListener(mDrawerToggle);
}
private void setupDrawerContent(NavigationView navigationView)
{navigationView.setNavigationItemSelectedListener(
new NavigationView.OnNavigationItemSelected
Listener() {
@Override
public boolean onNavigationItemSelected(MenuItem menuItem) {
menuItem.setChecked(true);
mDrawerLayout.closeDrawers();
return true;
}
});
```

拓展：当我们用手指滑动一个控件 View,从本质上来说就是移动一个 View, 改变其当前所处的位置,它的实现原理与动画效果的实现类似,都是通过不断的改变 View 的坐标来实现这个效果。以上内容讲解了 DrawerLayout 左 / 右滑抽屉布局效果。在 Android 中实现控件滑动的方法有许多种,扫描右侧二维码便可知晓。

技能点三　TabLayout

1　TabLayout 简介

自 2014 年 I/O 结束后,Google 在 Support Design 包中发布了一系列新的控件,其中就包括 TabLayout。在写项目时,通常在 ViewPager 上方放一个标签指示器与 ViewPager 进行联动。Tab 标签可以使用自定义 View,配合 ViewPager 和 Fragment 使用,TabLayout 可以帮助开发者即时打造一个滑动标签页。

2　TabLayout 的属性及效果图

TabLayout 使用过程中常用的属性如表 4.4 所示。

表 4.4　TabLayout 常用属性

属性	含义
app:tabMode	参数可选 fixed 和 scrollable——fixed 是指固定个数，scrollable 是使其可以横行滚动
app:tabGravity	对齐方式，可选 fill 和 center（注：此两种属性值只有在 tabMode 设置为 fixed 的情况下有效）
setSelectedTabIndicatorHeight	设置被选中标签下方导航条的高度
setTabTextColors	设置标签的字体颜色，1 为未选中标签的字体颜色，2 为被选中标签的字体颜色
setTabMode	TabMode 有两个可选参数： MODE_FIXED 表示宽度始终是 TabLayout 控件指定的宽度，如果标签过多，那么就无限挤压控件 MODE_SCROLLABLE 表示每个标签都保持自身宽度，一旦标签过多，给标题栏提供支持横向滑动的功能
setTabGravity	此条属性必须配合 MODE_FIXED 使用，不然不起作用。TabGravity 有两个可选参数： GRAVITY_FILL 让每个标签平分 TabLayout 的全部宽度 GRAVITY_CENTER 让每个标签显示自身宽度，然后所有标签居中显示
tabIndicatorColor	tab 指示符颜色
tabSelectedTextColor	tab 被选中字体颜色
setSelectedTabIndicatorColor	设置被选中标签下方导航条颜色
tabTextAppearance	改变 tab 的字体大小

通过学习以上属性，实现 Tab 标签功能，运行效果如图 4.11 所示。

图 4.11　Tab 效果图

3　TabLayout 的实现步骤

在应用的 build.gradle 中添加 support.design 支持库（注意：必须和 v7 包的版本相同）。具体代码如 CORE0409 所示。

代码 CORE0409　添加支持库
compile 'com.android.support:design:25.3.1'

创建布局文件 activity_short_tab，在布局文件中添加 TabLayout 及 ViewPager。具体代码如 CORE0410 所示。

代码 CORE0410　添加布局文件
<LinearLayout xmlns:android="http://schemas.android.com/apk/res/android" xmlns:app="http://schemas.android.com/apk/res-auto" xmlns:tools="http://schemas.android.com/tools" android:id="@+id/activity_main" android:layout_width="match_parent" android:layout_height="match_parent" android:orientation="vertical"

```xml
    tools:context="com.linwei.lw.tablayoutdemo.ui.activity.MainActivity">
<android.support.design.widget.TabLayout
    android:id="@+id/tab"
    android:layout_width="match_parent"
    android:layout_height="wrap_content"
    app:tabIndicatorColor="@color/colorAccent"
    app:tabIndicatorHeight="2dp"
    app:tabMode="fixed"
    app:tabSelectedTextColor="@color/colorAccent">
</android.support.design.widget.TabLayout>
<android.support.v4.view.ViewPager
    android:id="@+id/viewpager"
    android:layout_width="match_parent"
    android:layout_height="0dp"
    android:layout_weight="1">
</android.support.v4.view.ViewPager>
</LinearLayout>
```

定义一个 FragmentFactory 工厂类,生产 Fragment 对象,提高应用。具体代码如 CORE0411 所示。

代码 CORE0411　FragmentFactory 类

```java
// Fragment 工厂类。
public class FragmentFactory {
    private static HashMap<Integer, BaseFragment> mBaseFragments =
        new HashMap<Integer, BaseFragment>();
    public static BaseFragment createFragment(int pos) {
        BaseFragment baseFragment = mBaseFragments.get(pos);
        if (baseFragment == null) {
            switch (pos) {
                case 0:
                    baseFragment = new TopLineFragment();    // 头条
                    break;
                case 1:
                    baseFragment = new NewsFragment();    // 要闻
                    break;
                case 2:
```

```
                baseFragment = new EntertainmentFragment();// 娱乐
                break;
            case 3:
                baseFragment = new SportsFragment();    // 体育
                break;
        }
        mBaseFragments.put(pos, baseFragment);
    }
    return baseFragment;
}
```

定义一个 Fragment 的父类 BaseFragment。具体代码如 CORE0412 所示。

代码 CORE0412　BaseFragment 类

```
public abstract class BaseFragment extends Fragment {
    protected Context mContent;
    @Override
    public void onCreate(Bundle savedInstanceState) {
        super.onCreate(savedInstanceState);
        mContent = getContext();// 上下文。
    }
    @Override
    public View onCreateView(LayoutInflater inflater, ViewGroup container, Bundle savedInstanceState) {
        return initView();// 初始化布局。
    }
    @Override
    public void onActivityCreated(Bundle savedInstanceState) {
        super.onActivityCreated(savedInstanceState);
        loadData();// 初始化数据。
    }
    protected abstract void loadData();
    protected abstract View initView();
}
```

arrays.xml 中显示第一行数据。具体代码如 CORE0413 所示。

代码 CORE0413　arrays.xml 数据集合

```xml
<?xml version="1.0" encoding="utf-8"?>
<resources>
    <string-array name="tab_short_Title">
        <item>头条</item>
        <item>要闻</item>
        <item>娱乐</item>
        <item>体育</item>
    </string-array>
</resources>
```

ShortTabActivity 类，实现 TabLayout 和 ViewPager 的业务逻辑。具体代码如 CORE0414 所示。

代码 CORE0414　ShortTabActivity 类

```java
public class ShortTabActivity extends AppCompatActivity {
    private TabLayout mTab;                          // 初始化控件
    private ViewPager mViewPager;
    @Override
    protected void onCreate(Bundle savedInstanceState) {
        super.onCreate(savedInstanceState);
        setContentView(R.layout.activity_short_tab);
        initView();
        initData();
    }
    private void initData() {                        // 适配器分布标签
        ShortPagerAdapter adapter = new ShortPagerAdapter(
            getSupportFragmentManager());
        mViewPager.setAdapter(adapter);
        mTab.setupWithViewPager(mViewPager);
    }
    private void initView() {
        mTab = (TabLayout) findViewById(R.id.tab);
        mViewPager = (ViewPager) findViewById(R.id.viewpager);
    }
    private class ShortPagerAdapter extends FragmentPagerAdapter {
        public String[] mTilte;
        public ShortPagerAdapter(FragmentManager fm) {
```

```java
        super(fm);
        // 将标题添加至每个 fragment 上
        mTilte = getResources().getStringArray(R.array.tab_short_Title);
    }
    @Override
    public CharSequence getPageTitle(int position) {        // 得到标题
        return mTilte[position];
    }
    @Override
    public BaseFragment getItem(int position) {        // 得到点击的项
        BaseFragment fragment = FragmentFactory.createFragment(position);
        return fragment;
    }
    @Override
    public int getCount() {                            // 上下文
        return CommentUtils.TAB_SHORT_COUNT;
    }
}}
```

通过以上技能点的学习,掌握如何使用 DrawerLayout 实现侧滑菜单以及其他自定义控件的学习。实现侧滑菜单效果和新闻的预览。以下是实现侧滑菜单效果的具体步骤。

第一步:侧滑菜单的布局设计。具体代码如 CORE0415 所示。

代码 CORE0415　侧滑菜单布局

```xml
<android.support.v4.widget.DrawerLayout     xmlns:android="http://schemas.android.com/apk/res/android"
    xmlns:app="http://schemas.android.com/apk/res-auto"
    xmlns:tools="http://schemas.android.com/tools"
    android:id="@+id/drawer_layout"
    android:layout_width="match_parent"
    android:layout_height="match_parent"
    android:fitsSystemWindows="true"
    tools:context=".main.widget.MainActivity">
    <!-- 滚动效果 -->
    <android.support.design.widget.CoordinatorLayout
```

```xml
        android:id="@+id/main_content"
        android:layout_width="match_parent"
        android:layout_height="match_parent">
        <include
            android:id="@+id/appbar"
            layout="@layout/toolbar" />
        <FrameLayout
            android:id="@+id/frame_content"
            android:layout_width="match_parent"
            android:layout_height="match_parent"
            android:layout_below="@+id/appbar"
            android:scrollbars="none"
            android:elevation="5dp"
            app:layout_behavior="@string/appbar_scrolling_view_behavior" />
    </android.support.design.widget.CoordinatorLayout>
    <!-- NavigationView 为侧边抽屉栏 -->
    <android.support.design.widget.NavigationView
        android:id="@+id/navigation_view"
        android:layout_width="wrap_content"
        android:layout_height="match_parent"
        android:layout_gravity="start"
        app:headerLayout="@layout/navigation_header"
        app:menu="@menu/navigation_menu" />
</android.support.v4.widget.DrawerLayout>
```

第二步：实现侧滑菜单功能。具体代码如 CORE0416 所示。

代码 CORE0416　实现侧滑菜单功能

```java
protected void onCreate(Bundle savedInstanceState) {
super.onCreate(savedInstanceState);
setContentView(R.layout.activity_main);
mToolbar = (Toolbar) findViewById(R.id.toolbar);    // 调用 toolbar
setSupportActionBar(mToolbar);        //ToolBar 控件替代 ActionBar 控件
// 调用 findViewById() 方法得到了 Drawer Layout 的实例
mDrawerLayout = (DrawerLayout) findViewById(R.id.drawer_layout);
// 在左上角创建一个旋转按钮
mDrawerToggle = new ActionBarDrawerToggle(this,
mDrawerLayout, mToolbar, R.string.drawer_open, R.string.drawer_close);
```

```java
        mDrawerToggle.syncState();
        mDrawerLayout.setDrawerListener(mDrawerToggle);
        // 获取 Navigatioon View 的实例
        mNavigationView = (NavigationView) findViewById(R.id.navigation_view);
        setupDrawerContent(mNavigationView);
        mMainPresenter = new MainPresenterImpl(this);
        switch2News();
}
@Override
public boolean onCreateOptionsMenu(Menu menu) {        // 加载 toolbar 菜单文件
    getMenuInflater().inflate(R.menu.menu_main, menu);
    return true;
}
private void setupDrawerContent(NavigationView navigationView) {
    // 设置一个菜单项选中事件的监听器,当点击任意菜单项时,就会回调到
    //onNavigationItemSelected() 方法中
navigationView.setNavigationItemSelectedListener(new NavigationView.
    OnNavigationItemSelectedListener() {
                @Override
                public boolean onNavigationItemSelected(MenuItem menuItem) {
                    mMainPresenter.switchNavigation(menuItem.getItemId());
                    menuItem.setChecked(true);
                    mDrawerLayout.closeDrawers();
                    return true;
                }   });    }
```

第三步:对 mNavigationView 设置监听,实现了切换选项卡的效果。具体代码如 CORE0417 所示。

代码 CORE0417　设置监听效果

```java
@Override
public void switchNavigation(int id) {
    switch (id) {
        case R.id.navigation_item_news:
            mMainView.switch2News();
            break;
        case R.id.navigation_item_images:
            mMainView.switch2Images();
```

```
                break;
            case R.id.navigation_item_weather:
                mMainView.switch2Weather();
                break;
            case R.id.navigation_item_about:
                mMainView.switch2About();
                break;
            default:
                mMainView.switch2News();
                break;
        }
    }
}
```

第四步：创建四个 Fragment 来实现四个不同界面且可以跳转到主界面。具体代码如 CORE0418 所示。

代码 CORE0418　实现 Fragment 的切换界面

```
@Override
public void switch2News() {          // 进行切换界面
    getSupportFragmentManager().beginTransaction().replace(
    R.id.frame_content,new NewsFragment()).commit();
    mToolbar.setTitle(R.string.navigation_news);
}
@Override
public void switch2Images() {
    getSupportFragmentManager().beginTransaction().replace(
    R.id.frame_content,new ImageFragment()).commit();
    mToolbar.setTitle(R.string.navigation_images);
}
@Override
public void switch2Weather() {
    getSupportFragmentManager().beginTransaction().replace(
    R.id.frame_content,new WeatherFragment()).commit();
    mToolbar.setTitle(R.string.navigation_weather);
}
@Override
public void switch2About() {
    getSupportFragmentManager().beginTransaction().replace(
    R.id.frame_content,new AboutFragment()).commit();
```

mToolbar.setTitle(R.string.navigation_about);
}
// 实例化了 ActionBar 开关，同时调用 syncState() 同步状态，
// 后面对 mNavigationView 设置了监听，实现了切换选项卡的效果。

实现效果如图 4.12 所示。

图 4.12 滑动菜单效果图

以下为实现新闻预览的具体步骤。

第一步：新闻界面的主要布局。具体代码如 CORE0419 所示。

代码 CORE0419　新闻界面布局

```
<android.support.design.widget.CoordinatorLayout
xmlns:android="http://schemas.android.com/apk/res/android"
xmlns:app="http://schemas.android.com/apk/res-auto"
android:layout_width="match_parent"
android:layout_height="match_parent">
<android.support.design.widget.AppBarLayout
    android:layout_width="match_parent"
    android:layout_height="wrap_content"
    android:theme="@style/ThemeOverlay.AppCompat.Dark.ActionBar">
    <android.support.design.widget.TabLayout
```

```xml
        android:id="@+id/tab_layout"
        android:layout_width="match_parent"
        android:layout_height="?attr/actionBarSize"
        app:tabIndicatorColor="@color/icons"/>
</android.support.design.widget.AppBarLayout>
<android.support.v4.view.ViewPager
    android:id="@+id/viewpager"
    android:layout_width="match_parent"
    android:layout_height="wrap_content"
    app:layout_behavior="@string/appbar_scrolling_view_behavior"/>
</android.support.design.widget.CoordinatorLayout>
```

第二步:与服务器进行连接并获取数据,具体代码详见物料排序手持端项目。这里需要添加 okhttp3.2.0.jar 和 okio-1.6.jar 两个 jar 包。下载地址:(http://square.github.io/okhttp/)。

第三步:在主界面中嵌套了 FragmentList 布局,实现如图 4.13 所示效果。具体代码如 CORE0420 所示。

代码 CORE0420　　FragmentList 布局

```xml
<?xml version="1.0" encoding="utf-8"?>
<android.support.v4.widget.SwipeRefreshLayout xmlns:android="http://schemas.android.com/apk/res/android"
    xmlns:tools="http://schemas.android.com/tools"
    android:orientation="vertical"
    android:id="@+id/activity_main"
    android:layout_width="match_parent"
    android:layout_height="match_parent">
    <ListView
        android:layout_width="match_parent"
        android:layout_height="wrap_content"
        android:id="@+id/recyclerView">
    </ListView>
</android.support.v4.widget.SwipeRefreshLayout>
```

图 4.13 新闻界面效果图

第四步：在 FragmentList 中获取数据，实现如图 4.14 所示效果。具体代码如 CORE0421 所示。

```
代码 CORE0421    FragmentList 中获取数据
public class NewsListFragment extends Fragment {
// 初始化布局控件
ListView lv_titile;
List<String> l_title = new ArrayList<>();
List<String> l_img = new ArrayList<>();
List<String> l_content = new ArrayList<>();
okhttp okhttp=new okhttp();
private static final String TAG = "NewsListFragment";
// 初始化视图
private SwipeRefreshLayout mSwipeRefreshWidget;
private LinearLayoutManager mLayoutManager;
private List<News> mData;
private int mType = NEWS_TYPE_TOP;
private int pageIndex = 0;
String  url0="http://192.168.2.112:8080/SSMDemo/users/pro3",
        url1="http://192.168.2.112:8080/SSMDemo/users/pro25",
        url2="http://192.168.2.112:8080/SSMDemo/users/pro26",
```

```java
        url3="http://192.168.2.112:8080/SSMDemo/users/pro27";
    private NewsAdapter mAdapter;
    View view;
    // 重新加载出 View 以外的所有控件
    @Override
    public void onCreate(@Nullable Bundle savedInstanceState) {
        super.onCreate(savedInstanceState);
        mType = getArguments().getInt("type");
    }
    @Nullable
    @Override
    public View onCreateView(LayoutInflater inflater, ViewGroup container, Bundle savedInstanceState) {
        //view 显示 fragmentList 中的所有控件
        view = inflater.inflate(R.layout.fragment_newslist, container, false);
        mSwipeRefreshWidget = (SwipeRefreshLayout) view.findViewById(
                R.id.swipe_refresh_widget);
        mLayoutManager = new LinearLayoutManager(getActivity());
        onTakePhoto();
        initview();
        return view;
    }

    Handler handler=new Handler(){            //handler 数据处理
        @Override
        public void handleMessage(Message msg) {
            switch (msg.what){
                case 0:
                    String title=(String) msg.obj;
                    Log.d("title",title+"---------------");
                    if (title!=null) {
                        JsonData(title);
                    }
                    break;
            }        };
    // 将 fragmentlist 加载到 fragment 中
    public static NewsListFragment newInstance(int type) {
        Bundle args = new Bundle();
        NewsListFragment fragment = new NewsListFragment();
```

```java
        args.putInt("type", type);
        fragment.setArguments(args);
        return fragment;
    }
    private void getpost(String url) {            // 获取数据并解析 JSON
        File sdcache = getActivity().getExternalCacheDir();
        okhttp.initOkHttpClient(sdcache);
        try {
            RequestBody body = new FormBody.Builder().
            add("JSON", "{flag:object}").build();
            okhttp.postAsynHttp(url, body, "news");
            handler.postDelayed(new Runnable() {
                @Override
                public void run() {
                    String  str= Info.map.get("news");
                    Message msg = new Message();
                    msg.what = 0;
                    msg.obj = str;
                    handler.sendMessage(msg);
                }
            },1000);
        }catch (Exception o){
        };  }
    public void JsonData(String aa) {            //list 数据解析格式
        Gson gson = new Gson();                  // 调用 Gson
        Root rt = gson.fromJson(aa, Root.class);   // 调用 Root 实体类
        for (int i = 0; i < rt.getNews().size(); i++) {
            News data = rt.getNews().get(i);
            String tv_title = data.getTitle();
            String img_titlt = data.getImage();
            String tv_content = data.getContent();
            l_title.add(tv_title);
            l_img.add(img_titlt);
            l_content.add(tv_content);
        }
        ListAdapter adapter = new ListAdapter(getActivity(),
        l_title, l_img, l_content);
        lv_titile.setAdapter(adapter);
```

```
}
// 动态获取内存权限
public void onTakeStore () {
    if (Build.VERSION.SDK_INT>=23)    {
        int request= ContextCompat.checkSelfPermission(getActivity(),
        Manifest.permission. WRITE_EXTERNAL_STORAGE);
        // 缺少权限,进行权限申请
        if (request!= PackageManager.PERMISSION_GRANTED) {
            ActivityCompat.requestPermissions(getActivity(),new String[]{
            Manifest.permission.WRITE_EXTERNAL_STORAGE},123);
            return;
        }    else    {
            // 权限同意,不需要处理
            Toast.makeText(getActivity()," 权限同意 ",Toast.LENGTH_SHORT).show();
        }    }    else{
            // 低于 23 不需要特殊处理
    }    }}
```

图 4.14 新闻列表

第五步:点击或滑动 Viewpager 时跳转到 Fragment,实现如 4.15 所示效果。具体代码如 CORE0422 所示。

代码 CORE0422　Viewpager 点击或滑动时的事件

```java
private void initview(){
    lv_titile = (ListView) view.findViewById(R.id.recyclerView);
        // 点击 ViewPager 时，发生的数据请求
        switch (mType) {
            case NEWS_TYPE_TOP:
                getpost(url0);
// 点击 listview 后的界面跳转和数据传递
lv_titile.setOnItemClickListener(new AdapterView.OnItemClickListener() { @Override
public void onItemClick(AdapterView<?> adapterView, View view, int i, long l){
                    Intent intent = new Intent();
                    intent.putExtra("url", url0);
                    intent.putExtra("num", i);
                    intent.setClass(getActivity(), GetActivity.class);
                    startActivity(intent);
                }                               });
                break;
            case NEWS_TYPE_NBA:
                getpost(url2);
lv_titile.setOnItemClickListener(new AdapterView.OnItemClickListener() {
@Override
  public void onItemClick(AdapterView<?> adapterView, View view, int i, long l){
                    Intent intent = new Intent();
                    intent.putExtra("url", url2);
                    intent.putExtra("num", i);
                    intent.setClass(getActivity(), GetActivity.class);
                    startActivity(intent);
                }                               });
                break;
            case NEWS_TYPE_CARS:
                getpost(url3);
lv_titile.setOnItemClickListener(new AdapterView.OnItemClickListener() {
@Override
public void onItemClick(AdapterView<?> adapterView, View view, int i, long l){
                    Intent intent = new Intent();
                    intent.putExtra("url", url3);
                    intent.putExtra("num", i);
```

项目二　新闻天下

```
                              intent.setClass(getActivity(), GetActivity.class);
                              startActivity(intent);
                    }                              });
               break;
          case NEWS_TYPE_JOKES:
               getpost(url1);
lv_titile.setOnItemClickListener(new AdapterView.OnItemClickListener() {
Override
public void onItemClick(AdapterView<?> adapterView, View view, int i, long l) {
                    Intent intent = new Intent();
                    intent.putExtra("url", url1);
                    intent.putExtra("num", i);
                    intent.setClass(getActivity(), GetActivity.class);
                    startActivity(intent);
               }                              });
               break;          }    }
```

图 4.15　滑动后效果

第六步：点击 Listview 后跳转到的界面，将数据重新排列，实现如图 4.16 所示效果。具体代码如 CORE0423 所示。这里需要添加 gson-2.7.jar 包。

代码 CORE0423 Listview 跳转后的数据排列

```java
public class GetActivity extends AppCompatActivity {
// 初始化布局控件
okhttp okhttp=new okhttp();
TextView tv_get,tv_getting;
ImageView img_get;
// 定义参数
String url;
int num;
private AsyncImageLoader imageLoader;
@Override
protected void onCreate(Bundle savedInstanceState) {
    super.onCreate(savedInstanceState);
    setContentView(R.layout.activity_get);
    initview();
// 添加图片缓存方式
MemoryCache mcache = new MemoryCache();// 内存缓存
File sdCard = android.os.Environment.getExternalStorageDirectory();// 获得 SD 卡
File cacheDir = new File(sdCard, "jereh_cache");// 缓存根目录
FileCache fcache = new FileCache(this, cacheDir, "news_img");// 文件缓存
imageLoader = new AsyncImageLoader(this, mcache, fcache);
}
//handler 处理数据
Handler handler=new Handler(){
    @Override
    public void handleMessage(Message msg) {
        switch (msg.what){
            case 0:
                String strs= (String) msg.obj;
                JsonData(strs);
                break;
        }
    };
private void initview() {    // 找到布局中存在的控件
    tv_get= (TextView) findViewById(R.id.tv_get);
    tv_getting= (TextView) findViewById(R.id.tv_getting);
    img_get= (ImageView) findViewById(R.id.img_get);
    // 接受传递过来的参数
```

```java
        Intent intent=getIntent();
        url=intent.getStringExtra("url");
        num=intent.getIntExtra("num",0);
        getpost(url);
    }
//okhttp 请求服务器数据
private void getpost(String url) {
        File sdcache = this.getExternalCacheDir();
        okhttp.initOkHttpClient(sdcache);
        try {
RequestBody body = new FormBody.Builder().add("JSON", "{flag:object}").build();
            okhttp.postAsynHttp(url, body, "news");
            handler.postDelayed(new Runnable() {
                @Override
                public void run() {
                    String  str= Info.map.get("news");
                    Message msg = new Message();
                    msg.what = 0;
                    msg.obj = str;
                    handler.sendMessage(msg);
                }            },500);
    }catch (Exception o){
        };    }
// 解析获取的数据并显示
public void JsonData(String aa) {
    Gson gson = new Gson();
    Root rt = gson.fromJson(aa, Root.class);
        News data = rt.getNews().get(num);
        String tv_title = data.getTitle();
        String img_titlt = data.getImage();
        String tv_content = data.getContent();
    tv_get.setText(tv_title);
    Bitmap Bitmap = imageLoader.loadBitmap(img_get,img_titlt);
    img_get.setImageBitmap(Bitmap);
    tv_getting.setText(tv_content);
} }
```

图 4.16 新闻详情

第七步：设计新闻主界面及控件位置。具体代码如 CORE0424 所示。

代码 CORE0424   主界面的设计及控件

public class NewsFragment extends Fragment {
//ViewPage 的小标题
public static final int NEWS_TYPE_TOP = 0;
public static final int NEWS_TYPE_NBA = 1;
public static final int NEWS_TYPE_CARS = 2;
public static final int NEWS_TYPE_JOKES = 3;
// 初始化视图
private TabLayout mTablayout;
private ViewPager mViewPager;
@Nullable
@Override
public View onCreateView(LayoutInflater inflater, ViewGroup container,
　　Bundle savedInstanceState) {
　　// 获取主界面的所有控件
　　View view = inflater.inflate(R.layout.fragment_news, null);
　　mTablayout = (TabLayout) view.findViewById(R.id.tab_layout);
　　mViewPager = (ViewPager) view.findViewById(R.id.viewpager);
　　mViewPager.setOffscreenPageLimit(3);

```java
        setupViewPager(mViewPager);
        mTablayout.addTab(mTablayout.newTab().setText(R.string.top));
        mTablayout.addTab(mTablayout.newTab().setText(R.string.nba));
        mTablayout.addTab(mTablayout.newTab().setText(R.string.cars));
        mTablayout.addTab(mTablayout.newTab().setText(R.string.jokes));
        mTablayout.setupWithViewPager(mViewPager);
        return view;
    }
    private void setupViewPager(ViewPager mViewPager) {
        //Fragment 中嵌套使用 Fragment 一定要使用 getChildFragmentManager(),
        MyPagerAdapter adapter = new MyPagerAdapter(getChildFragmentManager());
        adapter.addFragment(NewsListFragment.newInstance(NEWS_TYPE_TOP), getString(R.string.top));
        adapter.addFragment(NewsListFragment.newInstance(NEWS_TYPE_NBA), getString(R.string.nba));
        adapter.addFragment(NewsListFragment.newInstance(NEWS_TYPE_CARS), getString(R.string.cars));
        adapter.addFragment(NewsListFragment.newInstance(NEWS_TYPE_JOKES), getString(R.string.jokes));
        mViewPager.setAdapter(adapter);
    }
    // 适配器填充 Fragment
    public static class MyPagerAdapter extends FragmentPagerAdapter {
        private final List<Fragment> mFragments = new ArrayList<>();
        private final List<String> mFragmentTitles = new ArrayList<>();
        public MyPagerAdapter(FragmentManager fm) {
            super(fm);
        }
        // 添加 fragment 及标题名称
        public void addFragment(Fragment fragment, String title) {
            mFragments.add(fragment);
            mFragmentTitles.add(title);
        }
        // 返回 Item 的点击位置
        @Override
        public Fragment getItem(int position) {
            return mFragments.get(position);
```

```
        }
    // 返回当前填充的 fragment 的大小
    @Override
    public int getCount() {
        return mFragments.size();
    }
    // 返回当前 Viewpage 的点击位置
    @Override
    public CharSequence getPageTitle(int position) {
        return mFragmentTitles.get(position);
    }   }}
```

本模块介绍了新闻天下项目新闻阅读模块的实现,通过本模块的学习可以了解 Toolbar 的具体功能和使用方法,掌握 TabLayout 控件的属性和效果。学习完成之后能够实现新闻列表刷新和查看新闻详情的功能。

虽然现在的标题栏是使用 Toolbar 来编写的,但看上去和传统的 ActionBar 没有什么不同,在 Material Design 中标题栏不只是固定的,也可通过自己的喜好随意制定标题栏的样式,其中就有一种可折叠式标题,是借助 Collapsing ToolbarLayout 来实现的。

技能扩展——ToolbarLayout

1　Collapsing ToolbarLayout 简介

Collapsing ToolbarLayout 是作用于 Toolbar 基础之上的布局,是由 Design Support 库提供的。Collapsing ToolbarLayout 可以让 Toolbar 的效果变得更加丰富,不仅展现一个标题栏,而是能够实现非常华丽的效果。Collapsing ToolbarLayout 是不能独立存在的。CollapsingToolbarLayout 一般作为 CoordinatorLayout 的子元素出现,另外一个控件 AppBarLayout 也是 Design 库的控件,作用是把其所有子元素当做一个 AppBar 来使用。

2　Collapsing ToolbarLayout 属性及效果图

Collasing ToolbarLayout 在使用过程中常用的属性如表 4.5 所示。

表 4.5　Collasing ToolbarLayout 常用属性

属性	含义
contentScrim	当 Toolbar 收缩到一定程度时的所展现的主体颜色
title	当 titleEnable 设置为 true 的时候,在 toolbar 展开的时候,显示大标题, toolbar 收缩时,显示为 toolbar 上面的小标题
scrimAnimationDuration	该属性控制 toolbar 收缩时,颜色变化的动画持续时间。即颜色变为 contentScrim 所指定的颜色进行的动画所需要的时间
expandedTitleGravity	指定 toolbar 展开时,title 所在的位置。类似的还有 expandedTitleMargin、collapsedTitleGravity 这些属性
collapsedTitleTextAppearance	指定 toolbar 收缩时,标题字体的样式,类似的还有 expandedTitleAppearance
scroll	所有想要滑动的控件都要设置这个标志位
enterAlways	设置了该标志位后,若 View 已经滑出屏幕,此时手指向下滑,View 会立刻出现,这是另一种使用场景
enterAlwaysCollapsed	设置了 minHeight,同时设置了该标志位的话,view 会以最小高度进度屏幕,当滑动控件滑动到顶部的时候才会拓展为完整的高度
exitUntilCollapsed	向上滑动时收缩当前 View。但 view 可以被固定在顶部
pin	有该标志位的 View 在界面滚动的过程中会一直停留在顶部,比如 Toolbar 可以被固定在顶部
parallax	有该标志位的 View 表示能和界面同时滚动

通过学习以上属性,实现 Collapsing ToolbarLayout 功能,运行效果如图 4.17 所示。

图 4.17　Collapsing ToolbarLayout 效果图

3 Collapsing ToolbarLayout 的实现步骤

在 budil.gradle 中需要添加 DesignSupport 库和 Cardview 库。具体代码如 CORE0425 所示。

代码 CORE0425 配置 Gradle 文件
compile'com.android.support:cardview-v7:25.3.1'

使用 CoordinatorLayout 来作为最外层的布局实现标题栏部分。具体代码如 CORE0426 所示。

代码 CORE0426 添加最外层布局

`<android.support.design.widget.CoordinatorLayout`
`xmlns:app="http://schemas.android.com/apk/res-auto"`
`android:layout_width="match_parent"`
`android:layout_height="match_parent">`
`<!-- 标题栏 -->`
`<android.support.design.widget.AppBarLayout`
`android:layout_width="match_parent"`
`android:layout_height="250dp">`
`</android.support.design.widget.AppBarLayout>`
`</android.support.design.widget.CoordinatorLayout>`

在 AppBarLayout 中嵌套一个 Collapsing ToolbarLayou,定义标题栏的具体内容,设置收缩时标题的颜色和展开时标题的颜色等。具体代码如 CORE0427 所示。

代码 CORE0427 添加 Collapsing ToolbarLayout

`<android.support.design.widget.CollapsingToolbarLayout`
`android:id="@+id/collapsing_toolbar"`
`android:layout_width="match_parent"`
`android:layout_height="match_parent"`
`android:theme="@style/ThemeOverlay.AppCompat.Dark.ActionBar"`
`app:contentScrim="?attr/colorPrimary"`
`app:layout_scrollFlags="scroll|exitUntilCollapsed">`
`<android.support.v7.widget.Toolbar`
`    android:id="@+id/toolbar"`
`    android:layout_width="match_parent"`
`    android:layout_height="?attr/actionBarSize"`
`    app:layout_collapseMode="pin"></android.support.v7.widget.Toolbar>`
`</android.support.design.widget.CollapsingToolbarLayout>`

使用 NestedScrollView 滚动的方式来实现内容详情的部分。具体代码如 CORE0428 所示。

代码 CORE0428　实现内容详情部分

```
<android.support.v4.widget.NestedScrollView
    android:layout_width="match_parent"
    android:layout_height="match_parent"
    app:layout_behavior="@string/appbar_scrolling_view_behavior">
    <LinearLayout
        android:orientation="vertical"
        android:layout_width="match_parent"
        android:layout_height="wrap_content">
        <android.support.v7.widget.CardView
            android:layout_width="match_parent"
            android:layout_height="wrap_content"
            android:layout_marginTop="15dp"
            android:layout_marginBottom="15dp"
            app:cardCornerRadius="4dp">
            <TextView
                android:id="@+id/content_text"
                android:layout_margin="10dp"
                android:layout_width="wrap_content"
                android:layout_height="wrap_content" />
        </android.support.v7.widget.CardView>
    </LinearLayout> </android.support.v4.widget.NestedScrollView>
```

填充界面上的内容,调用 setTitle() 方法将名字设置成当前界面的标题。具体代码如 CORE0429 所示。

代码 CORE0429　填充内容

```
public static final String Name="name";
public static final String Image_id="imagr_id";
@Override
protected void onCreate(Bundle savedInstanceState) {
    Intent intent=getIntent();
    String Name=intent.getStringExtra(Name);
    int Imageid=intent.getIntExtra(Image_id,0);
    CollapsingToolbarLayout collapsingToolbarLayout=(CollapsingToolbarLayout)
```

```
        findViewById(R.id.collapsing_toolbar);
        TextView textView=(TextView)findViewById(R.id.content_text);
        setSupportActionBar(toolbar);    // 将 toolbar 作为 ActionBar 显示
        ActionBar actionBar=getSupportActionBar();
        if (actionBar!= null){
            actionBar.setDisplayHomeAsUpEnabled(true);  // 默认一个返回箭头
        }
        collapsingToolbarLayout.setTitle(Name);  // 设置当前界面的标题
        String Content=generateContent(Name);  // 设置内容
        textView.setText(Content);
    }
    // 将名字循环 500 次,生成较长的字符串设置到 TextView 上
    private String generateContent(String name){
        StringBuilder Content=new StringBuilder();
        for (int i=0; i<500; i++) {
            Content.append(Name);
        }
        return Content.toString();
    }
```

Primary	主	Design	设计
Action	行动	Subtitle	小标题
Logo	商标	Gravity	重力
Mode	模式	Scrollable	滚动
Collapsing	折叠	Duration	持续时间

一、选择题

1. 下列关于 Toolbar 描述正确的选项是(　　)。

A. Toolbar 不仅继承了 ActionBar 的所有功能,而且使用更加灵活,可以自由的添加子控件

B. Toolbar 继承了 ActionBar 的少部分功能,而且使用更加灵活,可以自由的添加子控件

C. ActionBar 不仅继承了 Toolbar 的所有功能,而且使用更加灵活,可以自由的添加子控件

D. Toolbar 虽然继承了 ActionBar 的所有功能,而且使用更加灵活,但是不能自由的添加子控件

2. 关于 Toolbar 属性描述不正确的是(　　)。

A. colorPrimaryDark:状态栏的颜色(可用来实现沉浸效果)

B. android:textColorPrimary:Toolbar 中文字的颜色

C. colorPrimary:Toolbar 的背景颜色

D. app:navigationIcon:Toobar 中的小标题

3. 下列不是 Toolbar 元素的选项是(　　)。

A. 导航按钮

B. 导航窗格

C. 标题和子标题

D. 若干个自定义 View

4. "对齐方式,可选 fill 和 center"描述的是 TabLayout 的哪一个属性(　　)。

A. app:tabGravity

B. app:tabMode

C. setTabTextColors

D. setTabGravity

5. 关于 Collapsing ToolbarLayout 描述正确的是(　　)。

A. Collapsing ToolbarLayout 是作用于 Toolbar 基础之上的布局,是由 Design Support 库提供的

B. Collapsing ToolbarLayout 是作用于 TabLayout 基础之上的布局,是由 Design Support 库提供的

C. Collapsing ToolbarLayout 仅可以添加一个标题栏

D. Collapsing ToolbarLayout 可以独立存在

二、填空题

1. Toolbar 不仅继承了_____的所有功能,而且使用更加灵活,可以自由的添加_____。

2. Toolbar 包含的元素有_____、_____、_____、_____、_____。

3. Tab 标签可以使用自定义_____,配合着 ViewPager 和_____的使用,_____可以帮助开发者即时打造一个滑动标签页。

4. 在 TabLayout 中_____属性必须配合 MODE_FIXED 使用,不然不起作用。

5. 属性 app:tabMode 的参数可选_____和_____。

三、上机题

1. 编写代码实现列表加载功能。

2. 编写代码实现详情预览功能。

模块二 图片浏览

通过图片浏览模块的实现,了解滑动组件的使用方法,熟悉下拉加载图片的实现原理,掌握图片加载的流程以及方法,具备独立完成图片浏览模块的能力。
- 了解滑动组件的使用。
- 熟悉下拉刷新功能的实现。
- 掌握加载图片的方法。
- 具备数据解析能力。

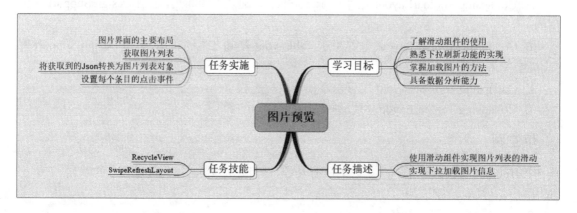

新闻天下提供给用户的功能不仅仅拘泥于浏览新闻,它是一款用于休闲娱乐的 App。新闻模块给用户提供时事政治,图片浏览模块可以给用户提供一种新的体验。本模块中图文以列表的形式显示,使用滑动组件实现图片列表的上下滑动。程序默认加载 20 条信息,供用户浏览更多信息。

【功能描述】

本模块将实现新闻天下项目中的图片浏览模块。
- 使用滑动组件实现图片列表的滑动。

● 实现下拉加载图片信息。

【基本框架】

基本框架如图 5.1 所示。

通过本模块的学习,将以上的框架图转换成图 5.2 所示效果。

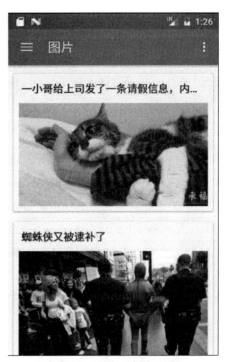

图 5.1　图片列表框架图　　　　图 5.2　图片列表效果图

技能点一　RecyclerView

ListView 已经不能满足大多数开发者的需求,而谷歌在 5.0 以后增加了新的列表控件 RecyclerView,它的性能及操作远远超过了 ListView 控件,下面将全面介绍 RecyclerView 的用法和操作步骤。

1　RecyclerView 简介

RecyclerView 是谷歌 V7 下新增的控件,用来替代 ListView 的使用,RecyclerView 不仅可以轻松实现和 ListView 同样的效果,还优化了 ListView 中存在的各种不足之处。通过设置

LayoutManager 快速实现 listview、gridview、瀑布流的效果,而且还可以设置横向和纵向显示,添加动画效果也非常简单(自带了 ItemAnimation),可以设置加载和移除时的动画,方便做出各种动态浏览的效果。

2 RecyclerView 的属性及效果图

RecyclerView 使用过程中常用的属性如表 5.1 所示。

表 5.1 RecyclerView 常用属性及含意

属性	含义
LinerLayoutManager	以垂直水平列表方式展示 Item
GridLayoutManager	以网格方式展示 Item
StaggeredGridLayoutManager	以瀑布流方式展示 Item
mRecyclerView.setLayoutManager(mLayoutManager);	设置布局管理器
mRecyclerView.setAdapter(mAdapter);	设置 adapter
mRecyclerView.setItemAnimator(new DefaultItemAnimator());	设置 Item 添加和移除的动画
mRecyclerView.addItemDecoration(mDividerItemDecoration);	设置 Item 之间间隔样式

通过学习以上属性,实现 Tab 标签功能,运行效果如图 5.3 和图 5.4 所示。

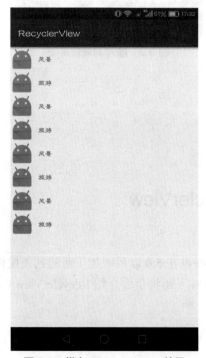

图 5.3 纵向 RecyclerView 效果

图 5.4 横向 RecyclerView 效果

3 RecyclerView 的实现步骤

（1）在应用的 build.gradle 中添加 support.design 支持库（注意：必须和 v7 包的版本相同）。具体代码如 CORE0501 所示。

代码 CORE0501 添加支持库

```
compile 'com.android.support:design:25.3.1'
```

（2）在 activity_main.xml 中添加 RecyclerView 布局。具体代码如 CORE0502 所示。

代码 CORE0502 添加 RecyclerView 布局

```xml
<LinearLayout
xmlns:android="http://schemas.android.com/apk/res/android"
android:layout_width="match_parent"
android:layout_height="match_parent" >
<android.support.v7.widget.RecyclerView
    android:id="@+id/recycle_view"
    android:layout_width="match_parent"
    android:layout_height="match_parent"></android.support.v7.widget.RecyclerView>
</LinearLayout>
```

（3）定义一个实体类，作为 RecyclerView 适配器的适配类型。具体代码如 CORE0503 所示。

代码 CORE0503　定义 fruit 实体类

```java
public class Fruit {
    private String name;
    private int imageId;
public  Fruit(String name,int imageId){
    this.name=name;
    this.imageId=imageId;
    }
    public String getName() {
        return name;
    }
    public void setName(String name) {
        this.name = name;
    }
    public int getImageId() {
```

```
        return imageId;
    }
    public void setImageId(int imageId) {
        this.imageId = imageId;
    }
}
```

（4）为RecyclerView的子项指定一个自定义布局。具体代码如CORE0504所示。

代码 CORE0504　自定义布局

```xml
<LinearLayout xmlns:android="http://schemas.android.com/apk/res/android"
    android:orientation="vertical"
    android:layout_width="100dp"
    android:layout_height="wrap_content">
    <ImageView
        android:id="@+id/fruit_image"
        android:layout_width="wrap_content"
        android:layout_height="wrap_content"
        <!--横向滚动 -->
        android:layout_gravity="center_horizontal"  />
    <TextView
        android:id="@+id/fruit_name"
        android:layout_width="wrap_content"
        android:layout_height="wrap_content"
        <!--横向滚动 -->
        android:layout_gravity="center_horizontal"
        android:layout_marginTop="10dp" />
</LinearLayout>
```

（5）创建FruitAdapter适配器继承RecyclerView.Adapter。具体代码如CORE0505所示。

代码 CORE0505　创建适配器

```java
public class FruitAdapter extends RecyclerView.Adapter<FruitAdapter.ViewHolder>{
    private List<Fruit> mFruitList;
    static class ViewHolder extends RecyclerView.ViewHolder{
        ImageView fruitImage;
        TextView fruitName;
        public ViewHolder(View view) {    // 定义内部类 ViewHolder
            super(view);    // 传入 View 参数
```

```
                fruitImage=(ImageView)view.findViewById(R.id.fruit_image);// 获取布局
                fruitName=(TextView)view.findViewById(R.id.fruit_name);
        } }
    public FruitAdapter(List<Fruit> fruitList) {
        mFruitList=fruitList;
    }
    @Override
    public ViewHolder onCreateViewHolder(ViewGroup parent, int viewType) {
        View view= LayoutInflater.from(parent.getContext()).inflate(
        R.layout.fruit_item,parent,false);       // 将 fruit_item 加载出来
        ViewHolder hodler=new ViewHolder(view);
        return hodler;
    }
    // 对 RecyclerView 子项的数据进行赋值
    @Override
    public void onBindViewHolder(ViewHolder holder, int position) {
        Fruit fruit=mFruitList.get(position);
        holder.fruitImage.setImageResource(fruit.getImageId());
        holder.fruitName.setText(fruit.getName());
    }
    @Override
    public int getItemCount() {
        return mFruitList.size();   // 返回数据源的长度
    }   }
```

（6）在 MainActivity 中使用 RecyclerView。具体代码如 CORE0506 所示。

代码 CORE0506　使用 RecyclerView

```
public class MainActivity extends AppCompatActivity {
private List<Fruit> fruitList = new ArrayList<>();
@Override
protected void onCreate(Bundle savedInstanceState) {
    // 获取 RecyclerVIew 的实例
    super.onCreate(savedInstanceState);
    setContentView(R.layout.activity_main);
    initFruits();
    RecyclerView recyclerView=(RecyclerView)findViewById(R.id.recycle_view);
    // 为线性布局
```

```
        LinearLayoutManager layoutManager=new LinearLayoutManager(this);
指定为横向布局
        layoutManager.setOrientation(LinearLayoutManager.HORIZONTAL);
        recyclerView.setLayoutManager(layoutManager);
        // 创建 FruitAdapter 实例,将数据传入到构建函数中
        FruitAdapter adapter=new FruitAdapter(fruitList);
        recyclerView.setAdapter(adapter); // 调用 setAdapter() 方法来完成适配器设置
    }
    private void initFruits() {   // 初始化数据
        for (int i = 0; i < 2; i++) {
            Fruit apple = new Fruit(" 风景 ",R.mipmap.ic_launcher);
            fruitList.add(apple);
            Fruit banana = new Fruit(" 旅游 ",R.mipmap.ic_launcher);
            fruitList.add(banana);
            Fruit orange= new Fruit(" 风景 ",R.mipmap.ic_launcher);
            fruitList.add(orange);
            Fruit pear = new Fruit(" 旅游 ",R.mipmap.ic_launcher);
            fruitList.add(pear);
        }    }    }
```

拓展:通过以上内容的学习,认识了新的列表控件 RecyclerView,列表控件用于显示数据集合,Android 中不是使用一种类型的控件管理显示和数据,而是将这两项功能分布用列表控件和适配器来实现。扫描右侧二维码,查看 Android 常用列表控件。

技能点二　SwipeRefreshLayout

通过安卓 5.0 之后的改版,SwipeRefreshLayout 已经将原本的线条刷新方式,变成了一个转动的圆环,加上原本控件的强大功能,一直被许多开发者所使用。下面将介绍 SwipeRefresh-Layout 的详细说明及其用法。

1　SwipeRefreshLayout 简介

SwipeRefreshLayout 是由 support-v4 库提供,用于实现下拉刷新功能。只接受需要刷新的子组件,通过 OnRefreshListener 设置监听,从监听里设置刷新需要获取的数据即可。

2　SwipeRefreshLayout 的属性及效果图

SwipeRefreshLayout 常用到的属性如表 5.2 所示。

表 5.2　SwipeRefreshLayout 常用属性

属性	含义
isRefreshing()	判断当前的状态是否是刷新状态
setColorSchemeResources(int... colorResIds)	设置下拉进度条的颜色主题，参数为可变参数
setOnRefreshListener(SwipeRefreshLayout.OnRefreshListener listener)	设置监听，需要重写 onRefresh() 方法，顶部下拉时会调用这个方法，在里面实现请求数据的逻辑，设置下拉进度条消失等
setProgressBackgroundColorSchemeResource(int colorRes)	设置下拉进度条的背景颜色，默认白色
setRefreshing(boolean refreshing)	设置刷新状态，true 表示正在刷新，false 表示取消刷新

使用 SwipeRefreshLayout 时要注意的事项：
- SwipeRefreshLayout 和 ScrollView 一样只能有一个子控件。
- setOnRefreshListener 设置监听刷新。
- setProgressBackgroundColor 设置刷新时圆形进度条的背景色。
- setColorSchemeResources 设置刷新时进度条颜色。
- setRefreshing 设置刷新状态。
- setSize 设置大小，只有 SwipeRefreshLayout.DEFAULT，SwipeRefreshLayout.LARGE。

通过学习以上属性，实现 SwipeRefreshLayout 功能，运行效果如图 5.5 所示；

图 5.5　SwipeRefreshLayout 效果图

3　SwipeRefreshLayout 的实现步骤

在 layout 中添加 SwipeRefreshLayout。具体代码如 CORE0507 所示。

代码 CORE0507　添加 SwipeRefreshLayout

<android.support.v4.widget.SwipeRefreshLayout
xmlns:tools="http://schemas.android.com/tools"
android:layout_width="match_parent"

```
    android:layout_height="match_parent"
    android:id="@+id/swipe_container">
<ScrollView
    android:layout_width="match_parent"
    android:layout_height="wrap_content">
</ScrollView></android.support.v4.widget.SwipeRefreshLayout>
```

在 Activity 中使用。具体代码如 CORE0508 所示。

代码 CORE0508　在 Activity 使用 SwipeRefreshLayout

```
private SwipeRefreshLayout  swipeRefreshLayout;
@Override
protected void onCreate(Bundle savedInstanceState) {
swipeRefreshLayout = (SwipeRefreshLayout)findViewById(R.id.swipe_container);
    // 设置下拉刷新进度条的颜色
    swipeRefreshLayout.setColorSchemeResources(R.color.colorAccent);
// 设置下拉刷新监听器
swipeRefreshLayout.setOnRefreshListener(
newSwipeRefreshLayout.OnRefreshListener() {
        @Override
        public void onRefresh() {
            new Handler().postDelayed(new Runnable() {
                @Override
                public void run() {
                    swipeRefreshLayout.setRefreshing(false);
    }     },6000);      });    }
```

通过以上技能点的学习,将实现本模块的图片预览功能。具体实现步骤如下所示。

第一步:使用 RecyclerView 设置图片界面的主要布局。具体代码如 CORE0509 所示。

代码 CORE0509　图片界面的主要布局

```
<LinearLayout
android:layout_width="match_parent"
android:layout_height="match_parent"
android:gravity="center_vertical"
```

```xml
    android:orientation="vertical"
    android:padding="8dp">
    <TextView
        android:id="@+id/tvTitle"
        android:layout_width="match_parent"
        android:layout_height="wrap_content"
        android:layout_marginBottom="8dp"
        android:padding="4dp"
        android:singleLine="true"
        android:textAppearance="@style/TextAppearance.AppCompat.Title"
        android:textColor="@color/primary_text"
        android:textSize="@dimen/sn_16sp" />
    <ImageView
        android:id="@+id/ivImage"
        android:layout_width="match_parent"
        android:layout_height="wrap_content"
        android:scaleType="centerCrop" />
</LinearLayout>
```

第二步：为 Recycler View 添加适配器。具体代码如 CORE0510 所示。

代码 CORE0510　添加适配器

```java
public class ImageAdapter extends RecyclerView.Adapter<ImageAdapter.ItemViewHolder> {
    private List<ImageBean> mData;
    private Context mContext;
    private OnItemClickListener mOnItemClickListener;
    public ImageAdapter(Context context) {
        this.mContext = context;
    }
    public void setmDate(List<ImageBean> data) {
        this.mData = data;
        this.notifyDataSetChanged();
    }
    // 将布局加载进来，
    @Override
    public ImageAdapter.ItemViewHolder onCreateViewHolder(
            ViewGroup parent, int viewType) {
```

```java
View v = LayoutInflater.from(parent.getContext()).inflate(
R.layout.item_image, parent, false);
// 创建 viewHolder 实例,将加载出来的布局传入构造函数中
ItemViewHolder vh = new ItemViewHolder(v);
// ViewHolder 的实例返回。
    return vh;
}
// 对数据进行赋值
@Override
public void onBindViewHolder(ImageAdapter.ItemViewHolder holder,
int position) {
    // 通过 position 参数得到当前项的实例
    ImageBean imageBean = mData.get(position);
        if(imageBean == null) {
        return;
    }
    holder.mTitle.setText(imageBean.getTitle());
    float scale = (float)imageBean.getWidth() / (float) mMaxWidth;
    int height = (int)(imageBean.getHeight() / scale);
    if(height > mMaxHeight) {
        height = mMaxHeight;
    }
    holder.mImage.setLayoutParams(new LinearLayout.LayoutParams(
    mMaxWidth, height));
    ImageLoaderUtils.display(mContext, holder.mImage,
    imageBean.getThumburl());
}
@Override
public int getItemCount() { // 返回数据源的长度
    if(mData == null) {
        return 0;
    }
    return mData.size();
}
public ImageBean getItem(int position) {
    return mData == null ? null : mData.get(position);
}
```

```java
public void setOnItemClickListener(OnItemClickListener onItemClickListener) {
    this.mOnItemClickListener = onItemClickListener;
}
public interface OnItemClickListener {
    void onItemClick(View view, int position);
}
```

第三步：获取图片列表，这里需要在 gradle 中添加"compile'com.github.bumptech.glide:-glide:3.6.1'"（图像缓存库）。具体代码如 CORE0511 所示。

代码 CORE0511　获取图片列表

```java
public class ImageModelImpl implements ImageModel {
    @Override
    public void loadImageList(final OnLoadImageListListener listener) {
        String url = Urls.IMAGES_URL;        // 提取网页中的图片
        OkHttpUtils.ResultCallback<String>loadNewsCallback=new
                OkHttpUtils.ResultCallback<String>() {
            @Override
            public void onSuccess(String response) {
                List<ImageBean>iamgeBeanList= ImageJsonUtils.readJsonImageBean(response);
                        // 将提取的结果放入 List 中
                        listener.onSuccess(iamgeBeanList);
            }
            @Override
            public void onFailure(Exception e) {
                listener.onFailure("load image list failure.", e);
            }
        };
        // 通过 okhttp 的 get 方法读取
        OkHttpUtils.get(url, loadNewsCallback);        }
public interface OnLoadImageListListener {
    void onSuccess(List<ImageBean> list);
    void onFailure(String msg, Exception e);
}   }
```

第四步：将获取到的 JSON 转换为图片列表对象，实现如图 5.6 所示效果。具体代码如 CORE0512 所示：

代码 CORE0512　　获取 JSON 并转换为图片列表

```java
public class ImageJsonUtils {
    private final static String TAG = "ImageJsonUtils";
    public static List<ImageBean> readJsonImageBeans(String res) {
        List<ImageBean> beans = new ArrayList<ImageBean>();
        try {
            JsonParser parser = new JsonParser();
            JsonArray jsonArray = parser.parse(res).getAsJsonArray();
            for (int i = 1; i < jsonArray.size(); i++) {
                JsonObject jo = jsonArray.get(i).getAsJsonObject();
                ImageBean news = JsonUtils.deserialize(jo, ImageBean.class);
                beans.add(news);
            }
        } catch (Exception e) {
            LogUtils.e(TAG, "readJsonImageBeans error", e);
        }
        return beans;
    }
}
```

图 5.6　图片列表

第五步：将获取到的信息填入相应的控件内。具体代码如 CORE0513 所示。

代码 CORE0513　显示图片标题和图片

```java
Public class ImageFragment extends Fragment implements ImageView,
SwipeRefreshLayout.
OnRefreshListener {
    private static final String TAG = "ImageFragment";
    private SwipeRefreshLayout mSwipeRefreshWidget;
    private RecyclerView mRecyclerView;
    private LinearLayoutManager mLayoutManager;
    private ImageAdapter mAdapter;
    private List<ImageBean> mData;
    private ImagePresenter mImagePresenter;
    @Override
    public void onCreate(@Nullable Bundle savedInstanceState) {
        super.onCreate(savedInstanceState);
        mImagePresenter = new ImagePresenterImpl(this);
    }
    @Nullable
    @Override
    public View onCreateView(LayoutInflater inflater, ViewGroup container,
    Bundle saved InstanceState){
        // 使用 view 找出布局中的所有控件
        View view = inflater.inflate(R.layout.fragment_image, null);
        mSwipeRefreshWidget = (SwipeRefreshLayout) view.findViewById
        (R.id.swipe_refresh_widget);
        mRecyclerView = (RecyclerView)view.findViewById(R.id.recycle_view);
        mRecyclerView.setHasFixedSize(true);
        mLayoutManager = new LinearLayoutManager(getActivity());
        mRecyclerView.setLayoutManager(mLayoutManager);
        mRecyclerView.setItemAnimator(new DefaultItemAnimator());
        // 加载适配器
        mAdapter = new ImageAdapter(getActivity().getApplicationContext());
        mRecyclerView.setAdapter(mAdapter);
        mRecyclerView.addOnScrollListener(mOnScrollListener);
        onRefresh();
        return view;
    }
    private RecyclerView.OnScrollListener mOnScrollListener =
```

```java
new RecyclerView.OnScroll
Listener() {
    private int lastVisibleItem;
    @Override
    public void onScrolled(RecyclerView recyclerView, int dx, int dy) {
        super.onScrolled(recyclerView, dx, dy);
        lastVisibleItem = mLayoutManager.findLastVisibleItemPosition();
    }
    @Override
    public void onScrollStateChanged(RecyclerView recyclerView, int newState) {
        super.onScrollStateChanged(recyclerView, newState);
        if (newState == RecyclerView.SCROLL_STATE_IDLE &&
            lastVisibleItem + 1 ==mAdapter.getItemCount() ) {
            // 加载更多
    Snackbar.make(getActivity().findViewById(R.id.drawer_layout),
        getString(R.string.image_hit), Snackbar.LENGTH_SHORT).show();
        } } };
@Override
public void onRefresh() {
    mImagePresenter.loadImageList();
}
@Override
public void addImages(List<ImageBean> list) {
    if(mData == null) {
        mData = new ArrayList<>();
    }
    mData.clear();
    mData.addAll(list);
    mAdapter.setmDate(mData);
}
@Override
public void showProgress() {
    mSwipeRefreshWidget.setRefreshing(true);
}
@Override
public void hideProgress() {
    mSwipeRefreshWidget.setRefreshing(false);
```

项目二　新闻天下

```
}
@Override
public void showLoadFailMsg() {    // 显示获取失败消息
    if (isAdded()) {
View view=getActivity()==null ? mRecyclerView.getRootView():
getActivity().findViewById(R.id.drawer_layout);
Snackbar.make(view, getString(R.string.load_fail),
Snackbar.LENGTH_SHORT).show();
    } } }
```

第六步：设置每个条目的点击事件。具体代码如 CORE0514 所示。

代码 CORE0514　　设置每个条目的点击事件

```
public class ItemViewHolder extends RecyclerView.ViewHolder implements
View.OnClickListener {
public TextView mTitle;
public ImageView mImage;
public ItemViewHolder(View v) {
    super(v);
    // 成员变量到初始化通过传入构造函数的 View 就实现
    mTitle = (TextView) v.findViewById(R.id.tvTitle);
    mImage = (ImageView) v.findViewById(R.id.ivImage);
    v.setOnClickListener(this);    // 设置每个条目的点击事件
}
@Override
public void onClick(View view) {
    if(mOnItemClickListener != null) {
        mOnItemClickListener.onItemClick(view, this.getPosition());
    } } }
```

本模块介绍了新闻天下项目图片浏览模块的实现,通过本模块的学习学会使用 RecyclerView 代替 ListView 实现浏览效果,掌握使用 SwipeRefreshLayout 实现下拉刷新的功能。学习之后能够实现滑动浏览图片的功能。

技能扩展——CoordinatorLayout

1 CoordinatorLayout 简介

CoordinatorLayout（协调者布局）作为协调一个或多个子控件的根布局，子控件使用 Behavior 和父控件或其他子控件进行交互。CoordinatorLayout 的使用核心就是 Behavior，使用 Behavior 来执行交互。

2 CoordinatorLayout 属性与效果图

CoordinatorLayout 的常用属性如表 5.3 所示。

表 5.3 CoordinatorLayout 的属性

属性	含义
Scrollqflag	和滚动联动都要设置这个标志
layout_behavio	把滚动的内容设置在 appbar 下面，并且滚动的内容会显示在（0,0）坐标

enterAlways: 跟随滚动视图的上下滚动，滚出屏幕。
● enterAlwaysCollapsed: 当滚动视图滚动到底时，View 只能以 minHeight 的高度滚入界面。
● exitUntilCollapsed: 跟随滚动视图的上下滚动，但滚出时会预留 minHeight 的高度，实际能滚动的距离为 (layout_height-minHeight)。
● snap: 根据滚动释放时的状态来自动执行完整的 enter 或者 exit 动画。

通过以上属性值的学习，实现浮动操作按钮与 Snackbar，Toolbar 的扩展与收缩等效果，如图 5.7 至图 5.9。

图 5.7　Snackbar 浮动按钮

图 5.8　Toolbar 的扩展

图 5.9　Toolbar 的收缩

3 CoordinatorLayout 实现步骤

在项目的 build.gradle 文件中，引入头像控件库和 CardView 库。

```
compile 'de.hdodenhof:circleimageview:1.3.0'
compile 'com.android.support:cardview-v7:23.1.0'
compile 'com.jakewharton:butterknife:7.0.1'
```

布局文件代码如 CORE0515 所示：

代码 CORE0515 布局代码

```xml
<?xml version="1.0" encoding="utf-8"?>
<android.support.design.widget.CoordinatorLayout
    xmlns:android="http://schemas.android.com/apk/res/android"
    xmlns:app="http://schemas.android.com/apk/res-auto"
    xmlns:tools="http://schemas.android.com/tools"
    android:layout_width="match_parent"
    android:layout_height="match_parent"
    android:fitsSystemWindows="false"
    tools:context=".MainActivity">
    <android.support.design.widget.AppBarLayout
        android:id="@+id/main_abl_app_bar"
        android:layout_width="match_parent"
        android:layout_height="wrap_content"
        android:theme="@style/ThemeOverlay.AppCompat.Dark.ActionBar">
        <android.support.design.widget.CollapsingToolbarLayout
            android:layout_width="match_parent"
            android:layout_height="450dp"
            app:layout_scrollFlags="scroll|exitUntilCollapsed|snap">
            <ImageView
                android:id="@+id/main_iv_placeholder"
                android:layout_width="match_parent"
                android:layout_height="300dp"
                android:scaleType="centerCrop"
                android:src="@drawable/large"
                app:layout_collapseMode="parallax" />
            <FrameLayout
                android:id="@+id/main_fl_title"
```

```xml
            android:layout_width="match_parent"
            android:layout_height="150dp"
            android:layout_gravity="bottom|center_horizontal"
            android:background="@color/colorPrimary"
            app:layout_collapseMode="parallax">
            <LinearLayout
                android:id="@+id/main_ll_title_container"
                android:layout_width="wrap_content"
                android:layout_height="wrap_content"
                android:layout_gravity="center"
                android:orientation="vertical">
                <TextView
                    android:layout_marginTop="@dimen/title_margin"
                    android:layout_width="wrap_content"
                    android:layout_height="wrap_content"
                    android:layout_gravity="center_horizontal"
                    android:gravity="bottom|center"
                    android:text="@string/person_name"
                    android:textColor="@android:color/white"
                    android:textSize="30sp"/>
                <TextView
                    android:layout_width="wrap_content"
                    android:layout_height="wrap_content"
                    android:layout_gravity="center_horizontal"
                    android:layout_marginTop="4dp"
                    android:text="@string/person_title"
                    android:textColor="@android:color/white" />
            </LinearLayout>
        </FrameLayout>
    </android.support.design.widget.CollapsingToolbarLayout>
</android.support.design.widget.AppBarLayout>
<android.support.v4.widget.NestedScrollView
    android:layout_width="match_parent"
    android:layout_height="match_parent"
    android:scrollbars="none"
    app:behavior_overlapTop="30dp"
    app:layout_behavior="@string/appbar_scrolling_view_behavior">
```

```xml
        <android.support.v7.widget.CardView
            android:layout_width="wrap_content"
            android:layout_height="wrap_content"
            android:layout_margin="8dp"
            app:cardElevation="8dp"
            app:contentPadding="16dp">
            <TextView
                android:layout_width="match_parent"
                android:layout_height="wrap_content"
                android:lineSpacingExtra="8dp"
                android:text="@string/person_intro" />
        </android.support.v7.widget.CardView>
</android.support.v4.widget.NestedScrollView>
<android.support.v7.widget.Toolbar
    android:id="@+id/main_tb_toolbar"
    android:layout_width="match_parent"
    android:layout_height="?attr/actionBarSize"
    android:background="@color/colorPrimary"
    app:layout_anchor="@id/main_fl_title"
    app:theme="@style/ThemeOverlay.AppCompat.Dark">
    <LinearLayout
        android:layout_width="wrap_content"
        android:layout_height="match_parent"
        android:orientation="horizontal">
        <Space
            android:layout_width="@dimen/image_final_width"
            android:layout_height="@dimen/image_final_width" />
        <TextView
            android:id="@+id/main_tv_title"
            android:layout_width="wrap_content"
            android:layout_height="match_parent"
            android:layout_marginLeft="8dp"
            android:gravity="center_vertical"
            android:text="@string/person_name"
            android:textColor="@android:color/white"
            android:textSize="20sp"
            android:visibility="invisible"/>
```

```
            </LinearLayout>
        </android.support.v7.widget.Toolbar>
        <de.hdodenhof.circleimageview.CircleImageView
            android:layout_width="@dimen/image_width"
            android:layout_height="@dimen/image_width"
            android:layout_gravity="center"
            android:src="@drawable/small"
            app:border_color="@android:color/white"
            app:border_width="2dp"
            app:layout_behavior=".AvatarImageBehavior" />
</android.support.design.widget.CoordinatorLayout>
```

在 onCreate 里面设置滑动逻辑，设置两个动画：监听 AppBar 的滑动，处理 Toolbar 和 Title 的显示，实现自动滑动效果。具体代码如 CORE0516 所示。

代码 CORE0516　滑动效果

```
@Override
    protected void onCreate(Bundle savedInstanceState) {
        super.onCreate(savedInstanceState);
        setContentView(R.layout.activity_main);
        ButterKnife.bind(this);
        mTbToolbar.setTitle("");
        // AppBar 的监听
        mAblAppBar.addOnOffsetChangedListener(new
        AppBarLayout.OnOffsetChangedListener() {
@Override
public void onOffsetChanged(AppBarLayout appBarLayout, int verticalOffset) {
            int maxScroll = appBarLayout.getTotalScrollRange();
            float percentage = (float) Math.abs(verticalOffset) / (float) maxScroll;
                handleAlphaOnTitle(percentage);
                handleToolbarTitleVisibility(percentage);
            }
        });
        initParallaxValues(); // 自动滑动效果 }
```

根据滑动百分比，设置 Title 和 Toolbar 的显示与消失，使用 Alpha 动画。

```
// 处理 ToolBar 的显示
    private void handleToolbarTitleVisibility(float percentage) {
```

```
            if (percentage >= PERCENTAGE_TO_SHOW_TITLE_AT_TOOLBAR) {
                if (!mIsTheTitleVisible) {
                    startAlphaAnimation(mTvToolbarTitle,
                        ALPHA_ANIMATIONS_DURATION, View.VISIBLE);
                    mIsTheTitleVisible = true;
                }
            } else {
                if (mIsTheTitleVisible) {
                    startAlphaAnimation(mTvToolbarTitle,
                        ALPHA_ANIMATIONS_DURATION, View.INVISIBLE);
                    mIsTheTitleVisible = false;
                }
            }
        }
        // 控制 Title 的显示
        private void handleAlphaOnTitle(float percentage) {
            if (percentage >= PERCENTAGE_TO_HIDE_TITLE_DETAILS) {
                if (mIsTheTitleContainerVisible) {
                    startAlphaAnimation(mLlTitleContainer,
                        ALPHA_ANIMATIONS_DURATION, View.INVISIBLE);
                    mIsTheTitleContainerVisible = false;
                }
            } else {
                if (!mIsTheTitleContainerVisible) {
                    startAlphaAnimation(mLlTitleContainer,
                        ALPHA_ANIMATIONS_DURATION, View.VISIBLE);
                    mIsTheTitleContainerVisible = true;
                }
            }
        }
        // 设置渐变的动画
        public static void startAlphaAnimation(View v, long duration, int visibility) {
            AlphaAnimation alphaAnimation = (visibility == View.VISIBLE)
                ? new AlphaAnimation(0f, 1f)
                : new AlphaAnimation(1f, 0f);
            alphaAnimation.setDuration(duration);
            alphaAnimation.setFillAfter(true);
            v.startAnimation(alphaAnimation);
        }
```

自动滑动动画,到一定比例时展开或关闭。具体代码如 CORE0517 所示。

代码 CORE0517　展开或关闭

```
// 设置自动滑动的动画效果
private void initParallaxValues() {
    CollapsingToolbarLayout.LayoutParams petDetailsLp =
        (CollapsingToolbarLayout.LayoutParams) mIvPlaceholder.getLayoutParams();
    CollapsingToolbarLayout.LayoutParams petBackgroundLp =
        (CollapsingToolbarLayout.LayoutParams) mFlTitleContainer.getLayoutParams();
    petDetailsLp.setParallaxMultiplier(0.9f);
    petBackgroundLp.setParallaxMultiplier(0.3f);
    mIvPlaceholder.setLayoutParams(petDetailsLp);
    mFlTitleContainer.setLayoutParams(petBackgroundLp);
}
```

到此联动效果就显示出来了，在 AppBar 中 Toolbar 和 Title 之间的关系，也符合 Material 的风格，给用户更多的体验。

Animation	动画	Recycler	回收站
Swipe	刷卡	Refresh	刷新
Scheme	格式	Crop	作物
Staggered	交错	Decoration	装饰
Default	默认	Visible	可见

一、选择题

1. 关于 RecyclerView 描述正确的是（　　）。

A. RecyclerView 不仅可以轻松实现和 ListView 同样的效果，还优化了 ListView 中存在的各种不足之处

B. RecyclerView 和 ListView 没区别

C. RecyclerView 功能不如 ListView 全面

D. ListView 不仅可以轻松实现和 RecyclerView 同样的效果，还优化了 RecyclerView 中存在的各种不足之处

2. 以下不是 RecyclerView 属性的选项是（　　）。

A. mRecyclerView.setLayoutManager(mLayoutManager);

B. mRecyclerView.setAdapter(mAdapter);

C. mRecyclerView.getItemAnimator(new DefaultItemAnimator());

D. mRecyclerView.addItemDecoration(mDividerItemDecoration);

3. "设置刷新状态，true 表示正在刷新，false 表示取消刷新"描述的是以下哪一个属性（　　）。

A. setRefreshing(boolean refreshing)

B. isRefreshing()

C. setColorSchemeResources(int... colorResIds)

D. setOnRefreshListener(SwipeRefreshLayout.OnRefreshListener listener)

4. 对属性"isRefreshing()"所描述的含义正确的是（　　）。

A. 判断当前的状态是否是刷新状态

B. 设置下拉进度条的颜色主题，参数为可变参数

C. 设置下拉进度条的背景颜色，默认白色

D. 设置监听，需要重写 onRefresh() 方法

5. 根据任务实施可知返回数据源长度的方法是（　　）。

A. getText()　　　　　　　　　　B. getItemCount()

C. getSharedPreferences()　　　　D. onRefresh()

二、填空题

1. 本项目的图片预览模块主要使用了_____和_____技术。

2. RecyclerView 通过设置 LayoutManager 来快速实现_____、_____、_____的效果。

3. SwipeRefreshLayout 只接受需要刷新的子组件，通过_____设置监听，从监听里设置即可。

4. 使用 SwipeRefreshLayout 时，_____设置刷新时进度条颜色。

5. 使用 SwipeRefreshLayout 时，setSize 设置大小有_____、_____。

三、上机题

1. 编写代码实现下拉刷新功能。

2. 编写代码实现 Tab 标签功能。

模块三　天气检测

通过天气检测模块的实现，了解获取经纬度并实现定位的功能，学习百度定位 API 接口的

使用,掌握 JSON 数据的解析过程,具有处理一般数据的能力。在任务实现过程中:
- 了解获取经纬度并实现定位的功能。
- 学习百度定位 API 接口的使用。
- 掌握 JSON 数据的简单分析。
- 具备处理一般数据的能力。

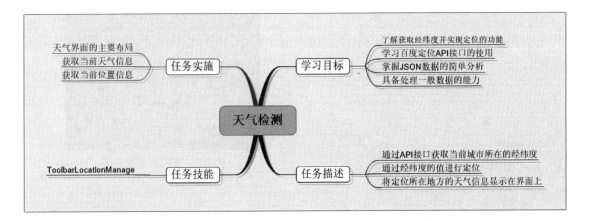

通过滑动侧边栏点击"天气"标签,进入到天气预测主界面。第一次进入主界面会提示用户是否开启 GPS 定位,接着出现一个刷新界面,当刷新完成后,会显示当前位置的一些天气信息(具体有时间、温度、风力、天气情况)。用户可以查看当前位置一周的天气信息,为需要出行的用户提供了极大的方便。

【功能描述】

本模块将实现新闻天下项目中的天气检测模块。
- 通过 API 接口获取当前城市所在的经纬度。
- 通过经纬度的值进行定位。
- 将定位所在地方的天气信息显示在界面上。

【基本框架】

基本框架如图 6.1 所示。
通过本模块的学习,将以上的框架图转换成图 6.2 所示效果。

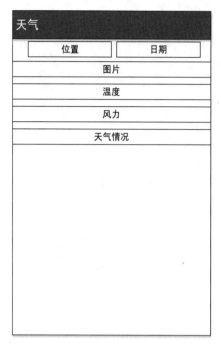
图 6.1 天气界面框架图

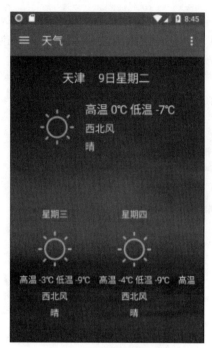
图 6.2 天气界面效果图

技能点一　LocationManager

  Android 软件开发中，常用到定位功能，尤其是依赖地理位置功能的应用。很多人喜欢使用百度地图、高德地图提供的 SDK。可是在只需要经纬度，或者城市，街道地址等信息时，并不需要提供预览地图界面在应用中。这时不需要使用百度地图或高德地图。这样做只会添加 APK 的体积。而在 LocationManager 中 Android API 给开发者提供的一些类就能够满足了，下面讲解怎样利用 LocationManager 获取经纬度。

1　LocationManager 简介

  Android 中的 LocationManager 用于显示获取移动设备所在的地理位置信息，其中包括经纬度。此外具有查询上一个已知位置、注册 / 注销来自某个 LocationProvider（提供定位功能的组件集合）的周期性位置更新以及注册 / 注销接近某个坐标时对一个已定义 Intent 的触发等功能。

2　LocationManager 相关方法

LocationManager 获取设备所在地理位置信息的功能是通过其内部的相关方法实现的，如表 6.1 是对 LocationManager 相关方法的介绍。

表 6.1　LocationManager 相关方法

方法	属性
addGpsStatusListener(GpsStatus.Listener listener)	添加 GPS 状态监听器
addNmeaListener(GpsStatus.NmeaListener listener)	添加监听器名称
addProximityAlert(double latitude, double longitude, float radius, long expiration, PendingIntent intent)	设置特定的位置提示，给定特定位置的经纬度坐标和半径
clearTestProviderEnabled(String provider)	在位置提供者中移除虚拟位置提供者
clearTestProviderLocation(String provider)	移除一些位置服务提供者提供的虚假位置
clearTestProviderStatus(String provider)	移除一些位置服提供者提供的虚拟状态
getAllProviders()	返回 List 包含名称的位置服务提供者
getBestProvider(Criteria criteria, boolean enabledOnly)	返回最符合标准的位置服务提供者的名称
getGpsStatus(GpsStatus status)	检测当前 GPS 设备的状态
getLastKnownLocation(String provider)	指定位置服务提供者名称返回该位置服务提供位置信息
getProvider(String name)	指定服务提供者名称返回实例
isProviderEnabled(String provider)	返回指定位置提供者是否启动
removeGpsStatusListener(GpsStatus.Listener listener)	移除 GPS 状态监听器

3　LocationManager 的用法

1. 获取 LocationManager 实例

```
LocationManager locationManager = (LocationManager)
    getSystemService(Context.LOCATION_SERVICE);
```

2. 选择位置提供器

```
String provider = LocationManager.NETWORK_PROVIDER;
```

- GPS_PROVIDER：GPS 定位的精准度比较高，但是非常耗电。
- NETWORK_PROVIDER：网络定位的精准度稍差，但耗电量比较少。
3. 将位置提供器传入到 getLastKnownLocation() 方法中，得到一个 Location 对象。

```
Location location = locationManager.getLastKnownLocation(provider);
location.getLatitude()    // 纬度
location.getLongitude()   // 经度
```

4. 用 requestLocationUpdates() 方法在设备位置发生改变的时候获取到最新的位置信息。

```
locationManager.requestLocationUpdates(LocationManager.GPS_PROVIDER, 5000, 10, new LocationListener() {
    @Override
    public void onStatusChanged(String provider, int status, Bundle extras) {
    }
    @Override
    public void onProviderEnabled(String provider) {
    }
    @Override
    public void onProviderDisabled(String provider) {
    }
    @Override
    public void onLocationChanged(Location location) {
    }
});
```

第一个参数：位置提供器的类型。
第二个参数：监听位置变化的时间间隔，以毫秒为单位。
第三个参数：监听位置变化的距离间隔，以米为单位。
第四个参数：LocationListener 监听器。

5. 权限的添加

```
<uses-permission android:name="android.permission.ACCESS_FINE_LOCATION"/>
<uses-permission android:name="android.permission.ACCESS_COARSE_LOCATION" />
```

由以上代码可知，LocationManager 每 5 秒钟会检测一次位置的变化情况，当移动距离超过 10 米的时候，就会调用 LocationListener 的 onLocationChanged() 方法，并把新的位置信息作为参数传入。

拓展：在我们的生活中，手机定位变得越来越重要。导航软件离不开定位、寻找好友位置离不开定位、推送天气预报离不开定位。其实这些都是通过卫星提供的 GPS 定位实现的。扫描右侧二维码，了解 GPS 定位实现原理解析。

通过上述技能点的介绍,将在下面介绍 LocationManager 在项目中的具体使用方法。最后达到定位,获取城市信息,并显示对应地区的天气情况的效果。具体实现步骤如下所示。

第一步:天气界面的主要布局。具体代码如 CORE0601 所示。

代码 CORE0601　天气界面的主要布局

`<FrameLayout`
`android:id="@+id/root_layout"`
`xmlns:android="http://schemas.android.com/apk/res/android"`
`android:layout_width="match_parent"`
`android:layout_height="match_parent"`
`android:orientation="vertical"`
`android:gravity="center_horizontal"`
`android:background="@drawable/biz_news_local_weather_bg_big">`
`<LinearLayout`
`    android:id="@+id/weather_layout"`
`    android:layout_width="match_parent"`
`    android:layout_height="match_parent"`
`    android:orientation="vertical"`
`    android:padding="12dp"`
`    android:visibility="gone"`
`    android:gravity="center_horizontal" >`
`    <LinearLayout`
`        android:layout_width="match_parent"`
`        android:layout_height="wrap_content"`
`        android:orientation="horizontal"`
`        android:gravity="center_horizontal"`
`        android:padding="12dp">`
`        <TextView`
`            android:id="@+id/city"`
`            style="@style/weacher_title"`
`            android:layout_width="wrap_content"`
`            android:layout_height="wrap_content"`
`            android:text=" 深圳 " />`
`        <TextView`
`            android:id="@+id/today"`

```xml
            style="@style/weacher_title"
            android:layout_width="wrap_content"
            android:layout_height="wrap_content"
            android:layout_marginLeft="24dp"
            android:text="2015年12月22日 星期二 " />
</LinearLayout>
<LinearLayout
        android:layout_width="match_parent"
        android:layout_height="wrap_content"
        android:gravity="center"
        android:padding="12dp"
        android:orientation="horizontal">
    <ImageView
            android:id="@+id/weatherImage"
            android:layout_width="wrap_content"
            android:layout_height="wrap_content"
            android:src="@drawable/biz_plugin_weather_qing" />
    <LinearLayout
            android:layout_width="wrap_content"
            android:layout_height="wrap_content"
            android:orientation="vertical"
            android:layout_marginLeft="8dp">
        <TextView
            android:id="@+id/weatherTemp"
            style="@style/weacher_title"
            android:layout_width="wrap_content"
            android:layout_height="wrap_content"
            android:padding="4dp"
            android:text="22℃ -28℃ " />
        <TextView
            android:id="@+id/wind"
            style="@style/weacher_temp"
            android:layout_width="wrap_content"
            android:layout_height="wrap_content"
            android:padding="4dp"
            android:text=" 微风 " />
        <TextView
            android:id="@+id/weather"
```

```xml
            style="@style/weacher_temp"
            android:layout_width="wrap_content"
            android:layout_height="wrap_content"
            android:padding="4dp"
            android:text=" 阵雨转阴 " />
        </LinearLayout>
    </LinearLayout>
    <HorizontalScrollView
        android:layout_width="match_parent"
        android:layout_height="match_parent" >
        <LinearLayout
            android:id="@+id/weather_content"
            android:layout_width="wrap_content"
            android:layout_height="match_parent"
            android:gravity="center"
            android:orientation="horizontal">
        </LinearLayout>
    </HorizontalScrollView>
    </LinearLayout>
    <ProgressBar
        android:id="@+id/progress"
        style="?android:attr/progressBarStyle"
        android:layout_width="wrap_content"
        android:layout_height="wrap_content"
        android:visibility="gone"
        android:layout_gravity="center"/>
</FrameLayout>
```

第二步：获取当前天气信息。具体代码如 CORE0602 所示。

代码 CORE0602　天气信息的读取

```java
public class WeatherModelImpl implements WeatherModel {
    private static final String TAG = "WeatherModelImpl";
    @Override
    // 获取当前天气信息
    public void loadWeatherData(String cityName, final LoadWeatherListener listener) {
        try { // 提取网页中的天气信息
            String url = Urls.WEATHER + URLEncoder.encode(cityName, "utf-8");
```

```java
            OkHttpUtils.ResultCallback<String> callback = new OkHttpUtils.ResultCallback<String>() {
                @Override
                public void onSuccess(String response) {
                    // 将提取的结果放入 List 中
                    List<WeatherBean> lists = WeatherJsonUtils.getWeatherInfo(response);
                    listener.onSuccess(lists);
                }
                @Override
                public void onFailure(Exception e) {
                    listener.onFailure("load weather data failure.", e);
                }
            };
            OkHttpUtils.get(url, callback);   // 通过 okhttp 的 get 方法读取
        } catch (UnsupportedEncodingException e) {
            LogUtils.e(TAG, "url encode error.", e);
        }
    }
```

第三步：获取当前位置信息。具体代码如 CORE0603 所示。

代码 CORE0603　当前位置信息

```java
public class WeatherModelImpl implements WeatherModel {
    private static final String TAG = "WeatherModelImpl";
    @Override
    // 在当前界面获取当前所在位置
    public void loadLocation(Context context, final LoadLocationListener listener) {
        // 获得 LocationManager 的引用
        LocationManager locationManager = (LocationManager)
                context.getSystemService(Context.LOCATION_SERVICE);
        if(Build.VERSION.SDK_INT >= Build.VERSION_CODES.M) {
            // 获取手机 GPS 使用权限
            if (context.checkSelfPermission(Manifest.permission.
                    ACCESS_FINE_LOCATION) !=
                    PackageManager.PERMISSION_GRANTED&&context.checkSelfPermission(
                    Manifest.permission.ACCESS_COARSE_LOCATION) !=
                    PackageManager.PERMISSION_GRANTED) {
                LogUtils.e(TAG, "location failure.");
                listener.onFailure("location failure.", null);
                return;
```

```java
        }        }
// 获取 Location
Location location=locationManager.getLastKnownLocation(
LocationManager.NETWORK_PROVIDER);
// 不为空显示地理位置经纬度
    if(location == null) {
        LogUtils.e(TAG, "location failure.");
        listener.onFailure("location failure.", null);
        return;
    }
    double latitude = location.getLatitude();      // 经度
    double longitude = location.getLongitude();    // 纬度
    String url = getLocationURL(latitude, longitude);  // 通过经纬度获取位置
    OkHttpUtils.ResultCallback<String> callback =
    new OkHttpUtils.ResultCallback<String>() {
    @Override
    public void onSuccess(String response) {
        // 定位 JSON 城市信息
        String city = WeatherJsonUtils.getCity(response);
                if(TextUtils.isEmpty(city)) {
            LogUtils.e(TAG, "load location info failure.");
            listener.onFailure("load location info failure.", null);
        } else {
            listener.onSuccess(city);   // 如果无法读取重新获取城市信息
        }        }
        @Override
        public void onFailure(Exception e) {
            LogUtils.e(TAG, "load location info failure.", e);
            listener.onFailure("load location info failure.", e);
        }        };
    OkHttpUtils.get(url, callback);
}
// 解析城市 JSON 串,获取城市信息
private String getLocationURL(double latitude, double longitude) {
    StringBuffer sb = new StringBuffer(Urls.INTERFACE_LOCATION);
    sb.append("?output=json").append("&referer=
    32D45CBEEC107315C553AD1131915D366EEF79B4");
    sb.append("&location=").append(latitude).append(",").append(longitude);
```

```
        LogUtils.d(TAG, sb.toString());
        return sb.toString();
    }
```

第四步：显示定位信息和天气信息。具体代码如 CORE0604 所示。

代码 CORE0604　显示定位信息和天气信息

```java
public class WeatherPresenterImpl implements WeatherPresenter,
WeatherModelImpl.LoadWeatherListener {
// 初始视图信息
private WeatherView mWeatherView;
private WeatherModel mWeatherModel;
private Context mContext;
public WeatherPresenterImpl(Context context, WeatherView weatherView) {
    this.mContext = context;
    this.mWeatherView = weatherView;
    mWeatherModel = new WeatherModelImpl();
}
@Override
// 下载天气信息到上下文
public void loadWeatherData() {
    mWeatherView.showProgress();
    if(!ToolsUtil.isNetworkAvailable(mContext)) {
        mWeatherView.hideProgress();
        mWeatherView.showErrorToast(" 无网络连接 ");
        return;
    }
    // 位置监听器
    WeatherModelImpl.LoadLocationListener listener = new
    WeatherModelImpl.LoadLocationListener() {
        @Override
        public void onSuccess(String cityName) {
            // 定位成功,获取定位城市天气预报
            mWeatherView.setCity(cityName);
            mWeatherModel.loadWeatherData(cityName, WeatherPresenterImpl.this);
        }
        @Override
        public void onFailure(String msg, Exception e) {
```

```
            mWeatherView.showErrorToast(" 定位失败 ");
            mWeatherView.setCity(" 深圳 ");
            mWeatherModel.loadWeatherData(" 深圳 ", WeatherPresenterImpl.this);
        }  };
    // 获取定位信息
    mWeatherModel.loadLocation(mContext, listener);
}
@Override
public void onSuccess(List<WeatherBean> list) {    // 获取这一周的天气信息
    if(list != null && list.size() > 0) {
        WeatherBean todayWeather = list.remove(0);
        mWeatherView.setToday(todayWeather.getDate());
        mWeatherView.setTemperature(todayWeather.getTemperature());
        mWeatherView.setWeather(todayWeather.getWeather());
        mWeatherView.setWind(todayWeather.getWind());
        mWeatherView.setWeatherImage(todayWeather.getImageRes());
    }
    // 获取的天气信息,填入视图中
    mWeatherView.setWeatherData(list);
    mWeatherView.hideProgress();
    mWeatherView.showWeatherLayout();
}
@Override
public void onFailure(String msg, Exception e) {
    mWeatherView.hideProgress();
    mWeatherView.showErrorToast(" 获取天气数据失败 ");
}}
```

第五步：将获取的信息填入相应的控件内，实现如图 6.3 所示效果。具体代码如 CORE0605 所示。

代码 CORE0605　　显示定位信息和天气信息
| ```
public class WeatherFragment extends Fragment implements WeatherView {
// 初始化视图信息和控件信息
private WeatherPresenter mWeatherPresenter;
private TextView mTodayTV;
private ImageView mTodayWeatherImage;
private TextView mTodayTemperatureTV;
``` |
```

```java
    private TextView mTodayWindTV;
    private TextView mTodayWeatherTV;
    private TextView mCityTV;
    private ProgressBar mProgressBar;
    private LinearLayout mWeatherLayout;
    private LinearLayout mWeatherContentLayout;
    private FrameLayout mRootLayout;
    @Override
    public void onCreate(Bundle savedInstanceState) {
        super.onCreate(savedInstanceState);
        mWeatherPresenter = new WeatherPresenterImpl(
        getActivity().getApplication(), this);
    }
    @Override
    public View onCreateView(LayoutInflater inflater, ViewGroup container, Bundle savedInstanceState) {
    // 使用 View 找出布局中的所有控件
    View view = inflater.inflate(R.layout.fragment_weather, null);
    mTodayTV = (TextView) view.findViewById(R.id.today);
    mTodayWeatherImage = (ImageView) view.findViewById(R.id.weatherImage);
    mTodayTemperatureTV = (TextView) view.findViewById(R.id.weatherTemp);
    mTodayWindTV = (TextView) view.findViewById(R.id.wind);
    mTodayWeatherTV = (TextView) view.findViewById(R.id.weather);
    mCityTV = (TextView)view.findViewById(R.id.city);
    mProgressBar = (ProgressBar) view.findViewById(R.id.progress);
    mWeatherLayout = (LinearLayout) view.findViewById(R.id.weather_layout)
    mWeatherContentLayout =(LinearLayout)view.findViewById(R.id.weather_content);
    mRootLayout = (FrameLayout) view.findViewById(R.id.root_layout);
        mWeatherPresenter.loadWeatherData();
        return view;
    }
    @Override
    public void showProgress() {                    // 显示滚动列表
        mProgressBar.setVisibility(View.VISIBLE);
    }
    @Override
    public void hideProgress() {                    // 隐藏滚动列表
        mProgressBar.setVisibility(View.GONE);
```

```java
}
@Override
public void showWeatherLayout() {              // 显示天气
    mWeatherLayout.setVisibility(View.VISIBLE);
}
@Override                                       // 显示城市信息
public void setCity(String city) {
    mCityTV.setText(city);
}
@Override                                       // 显示星期信息
public void setToday(String data) {
    mTodayTV.setText(data);
}
@Override                                       // 显示温度信息
public void setTemperature(String temperature) {
    mTodayTemperatureTV.setText(temperature);
}
@Override
public void setWind(String wind) {
    mTodayWindTV.setText(wind);
}
@Override                                       // 显示天气类型
public void setWeather(String weather) {
    mTodayWeatherTV.setText(weather);
}
@Override                                       // 显示天气图片
public void setWeatherImage(int res) {
    mTodayWeatherImage.setImageResource(res);
}
@Override
public void setWeatherData(List<WeatherBean> lists) {
    List<View> adapterList = new ArrayList<View>();
    // 将得到的数据信息重新加载在布局中
    for (WeatherBean weatherBean : lists) {
        View view = LayoutInflater.from(getActivity()).inflate(
    R.layout.item_weather, null, false);
        TextView dateTV = (TextView) view.findViewById(R.id.date);
        ImageView todayWeatherImage = (ImageView) view.findViewById(
```

```
            R.id.weatherImage);
        TextView todayTemperatureTV = (TextView) view.findViewById(
    R.id.weatherTemp);
        TextView todayWindTV = (TextView) view.findViewById(R.id.wind);
        TextView todayWeatherTV = (TextView) view.findViewById(
    R.id.weather);
        dateTV.setText(weatherBean.getWeek());
        todayTemperatureTV.setText(weatherBean.getTemperature());
        todayWindTV.setText(weatherBean.getWind());
        todayWeatherTV.setText(weatherBean.getWeather());
        todayWeatherImage.setImageResource(weatherBean.getImageRes());
        mWeatherContentLayout.addView(view);
        adapterList.add(view);
    }   }
@Override
// 显示出错信息
public void showErrorToast(String msg) {
Snackbar.make(getActivity().findViewById(R.id.drawer_layout), msg, Snackbar.LENGTH_SHORT).show();
    }}
```

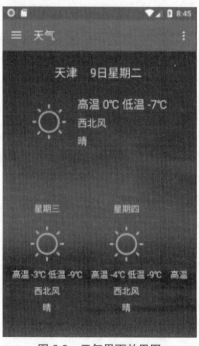

图 6.3　天气界面效果图

本模块介绍了新闻天下项目天气检测模块的实现,通过本模块的学习可以了解网络接口 API 的具体功能和使用方法,掌握 LocationManager 的用法。学习完成后能够实现天气信息的监控和天气详情的浏览。

技能扩展——LocationManage

LocationManager 也可以实现定位,但是其功能具有很大的局限性。如果要在应用中出现导航的功能,并在界面中出现地图,那么 LocationManager 是达不到要求的,所以开发者还需要借助百度地图提供的 SDK 来实现这样的功能。以下技能扩展,将简单的介绍百度地图 SDK 的使用方法。

1 Android 定位 SDK

百度地图 Android 定位 SDK 是为 Android 移动端应用提供的一套简单易用的定位服务接口,专注于为广大开发者提供最好的综合定位服务。通过使用百度定位 SDK,开发者可以轻松为应用程序实现智能、精准、高效的定位功能。

2 Android Studio 配置

第一步:打开 / 创建一个 Android 工程

根据开发者的实际使用情况,打开一个已有 Android 工程,或者新建一个 Android 工程。这里以新建一个 Android 工程为例讲解。

第二步:添加 SDK(jar+so)

下载 Android 定位 SDK 并解压,将 libs 中的 jar 和 so 放置到工程中相应的位置。注意,Android 定位 SDK 提供了多种 CPU 架构的 so 文件(jar 通用,只有一个),开发者可根据实际使用需求,放置所需 so 到对应的工程文件夹内。图 6.4 为 Android 定位 SDK 文件结构示意图:

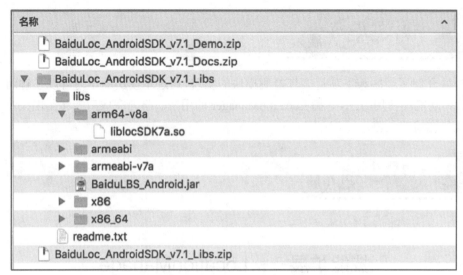

图 6.4　SDK 文件结构示意图

第三步：配置 build.gradle 文件

如图 6.5 所示，配置 build.gradle 文件，注意设置 sourceSets。

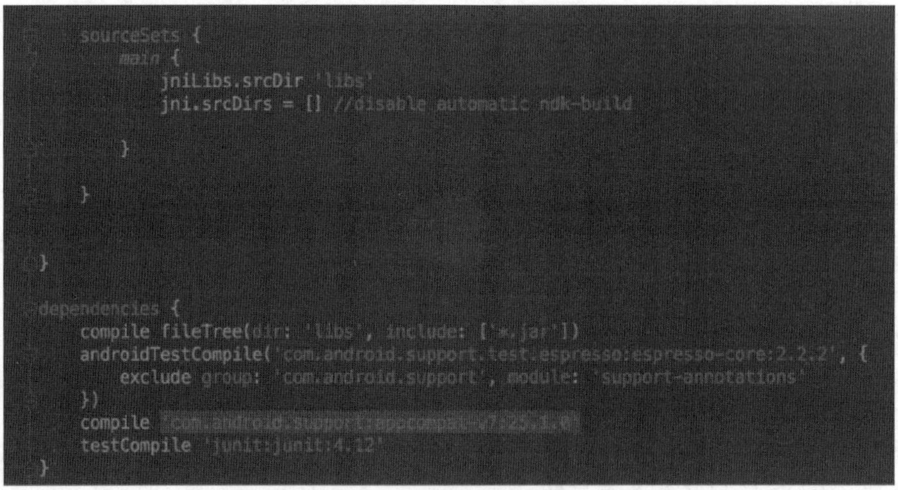

图 6.5　配置 build.gradle

第四步：添加 AK

Android 定位 SDK 自 v4.0 版本起，需要进行 AK 鉴权。开发者在使用 SDK 前，需完成 AK 申请，并在 AndroidManifest.xml 文件中，正确填写 AK。

在 Application 标签中增加如下代码：

```
<meta-data
    android:name="com.baidu.lbsapi.API_KEY"
    android:value=" 开发者申请的 AK" >
</meta-data>
```

如图 6.6 所示:

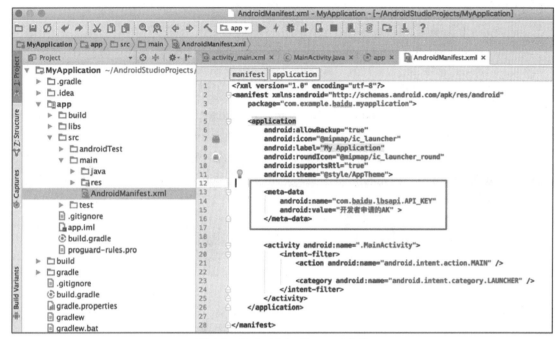

图 6.6　AndroidManifest.xml 文件

第五步:添加定位权限

使用定位 SDK,需在 Application 标签中声明 service 组件,每个 App 拥有自己单独的定位 service。具体代码如下所示:

<service   android:name="com.baidu.location.f"   android:enabled="true"   android:process=":remote"> </service>

除添加 service 组件外,使用定位 SDK 还需添加如下权限:

<!-- 这个权限用于进行网络定位 -->
<uses-permission   android:name="android.permission.ACCESS_COARSE_LOCATION"></uses-permission>
<!-- 这个权限用于访问 GPS 定位 -->
<uses-permission   android:name="android.permission.ACCESS_FINE_LOCATION"></uses-permission>
<!-- 用于访问 wifi 网络信息,wifi 信息会用于进行网络定位 -->
<uses-permission   android:name="android.permission.ACCESS_WIFI_STATE"></uses-permission>
<!-- 获取运营商信息,用于支持提供运营商信息相关的接口 -->

<uses-permission   android:name="android.permission.ACCESS_NETWORK_STATE"></uses-permission>

<!-- 这个权限用于获取wifi的获取权限,wifi信息会用来进行网络定位 -->

<uses-permission android:name="android.permission.CHANGE_WIFI_STATE"></uses-permission>

<!-- 用于读取手机当前的状态 -->

<uses-permission android:name="android.permission.READ_PHONE_STATE"></uses-permission>

<!-- 写入扩展存储,向扩展卡写入数据,用于写入离线定位数据 -->

<uses-permission   android:name="android.permission.WRITE_EXTERNAL_STORAGE"></uses-permission>

<!-- 访问网络,网络定位需要上网 -->

<uses-permission android:name="android.permission.INTERNET" />

<!-- SD卡读取权限,用户写入离线定位数据 -->

3 密钥申请方式

Key的申请地址为:http://lbsyun.baidu.com/apiconsole/key。申请的具体过程如下所示。

第一步:登录百度API控制台,登录会跳转到API控制台服务,具体如图6.7所示:

图6.7 百度API控制台

第二步:点击"创建应用",进入创建AK界面,输入应用名称,将应用类型改为:"Android SDK",如图6.8和图6.9所示:

图 6.8 创建 AK 界面

图 6.9 更改应用类型

第三步:在应用类型选择"Android SDK"后,需要配置应用的安全码,如图 6.10 所示:

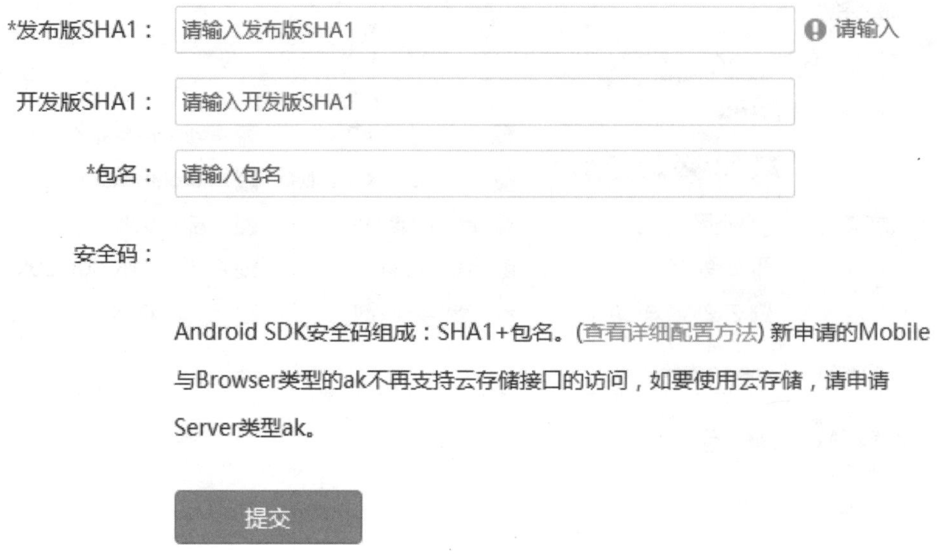

*发布版SHA1： [请输入发布版SHA1] ⓘ 请输入

开发版SHA1： [请输入开发版SHA1]

*包名： [请输入包名]

安全码：

Android SDK安全码组成：SHA1+包名。(查看详细配置方法)新申请的Mobile与Browser类型的ak不再支持云存储接口的访问，如要使用云存储，请申请Server类型ak。

[提交]

图 6.10　安全码

第四步：获取安全码。安全码的组成规则为：Android 签名证书的 sha1 值 +";"+packagename(即：数字签名 + 分号 + 包名)。

Android 签名证书的 sha1 值获取方式有两种：

第一种方法：使用 keytool

第 1 步：运行进入控制台，如图 6.11 所示。

图 6.11　控制台

第 2 步：定位到 .android 文件夹下，输入 cd.android，如图 6.12 所示。

项目二 新闻天下

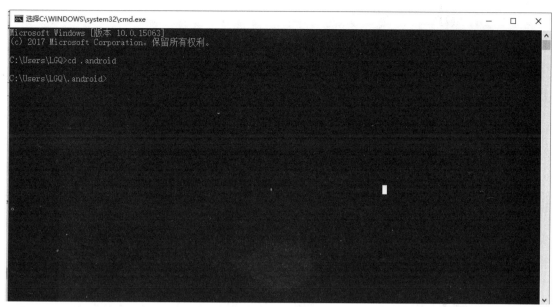

图 6.12 定位文件夹

第 3 步：输入 keytool -list -v -keystore debug.keystore，会得到三种指纹证书，选取 SHA1 类型的证书（密钥口令是 android），如图 6.13 所示。

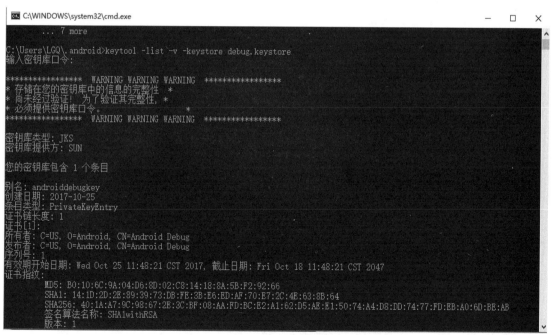

图 6.13 SHA1

第二种方法：在 Android Studio 中操作，如图 6.14 所示。

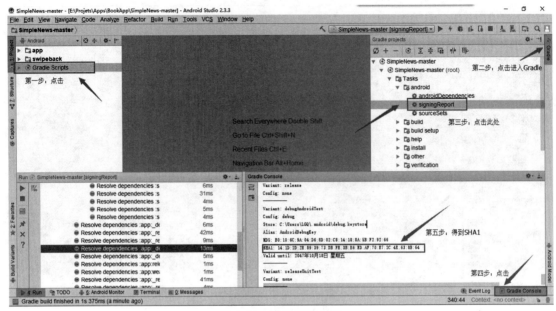

图 6.14 得到 SHA1 码

Location	方向	Latitude	经度
Longitude	维度	Change	改变
Listener	听	Weather	天气
Append	添加	Implement	实行
Visible	显示	Gone	过去的

一、选择题

1. 通过 LocationManager 的学习，选出获取纬度的方法（　　）。
   A. getLatitude()　　　　　　　　B. getLongitude()
   C. getSystemService()　　　　　D. getLocationURL()

2. LocationManager 所具有的功能描述不正确的是（　　）。
   A. 查询上一个已知位置
   B. 注册/注销来自某个 LocationProvider 的周期性的位置更新
   C. 注册/注销接近某个坐标时对一个已定义 Intent 的触发
   D. 帮助用户获取非常精确的位置信息

3. 位置提供器中网络定位精度稍差，耗电较少的是（　）。

A. GPS_PROVIDER

B. NETWORK_PROVIDER

C. Context.LOCATION_SERVICE

D. ACCESS_COARSE_LOCATION

4. 用于访问 GPS 定位的权限是（　）。

A. <uses-permission android:name="android.permission.ACCESS_FINE_LOCATION"></uses-permission>

B. <uses-permission android:name="android.permission.ACCESS_WIFI_STATE"></uses-permission>

C. <uses-permission android:name="android.permission.ACCESS_NETWORK_STATE"></uses-permission>

D. <uses-permission android:name="android.permission.CHANGE_WIFI_STATE"></uses-permission>

5. 在设备位置发生改变的时候获取到最新的位置信息的方法是（　　）。

A. requestLocationUpdates()　　　　B. getLastKnownLocation()

C. getSystemService()　　　　　　　D. getApplication()

二、填空题

1. Android 中的_____用于显示获取移动设备所在的地理位置信息，其中包括_____。

2. _____GPS 定位的精准度比较高，但是非常耗电。

3. requestLocationUpdates() 方法中第_____个参数是监听位置变化的时间间隔，以毫秒为单位。

4. requestLocationUpdates() 方法中参数 LocationManager.GPS_PROVIDER 代表的意思是_____。

5. LocationManager 中当移动距离超过 10 米时会调用 LocationListener 的_____方法。

三、上机题

1. 编写代码实现定位功能。

2. 配置 Android 定位 SDK。

项目三 微聊

模块一 系统及个人

通过系统设置和个人信息模块的实现，了解 Retrofit 的网络连接工具，学习使用简单的 MVP 框架来完成数据与视图之间的联系，具备将模块进行分工并简化代码的能力。在任务实施过程中：

- 了解 Retrofit 网络连接工具。
- 学习使用 MVP 框架。
- 掌握数据与视图之间的联系。
- 具备将模块进行分工并简化代码的能力。

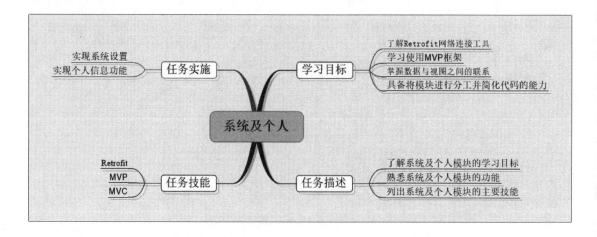

进入程序首先是一个闪屏界面,当闪屏界面结束后进入到程序的操作界面进行登录和注册。如果有账号则直接登录,无账号则需要进行注册。注册过程中需要通过手机短信验证码进行手机号的绑定。注册成功后,直接跳转至主界面。在主界面中登陆完成后可进行修改昵称、修改头像,可以生成用户自己的二维码图片,其他用户可通过扫描二维码进行好友添加等操作。

【功能描述】

本模块将实现本项目中的条形码扫描模块
- 闪屏界面的视图效果。
- 手机验证码验证登录。
- 修改个人信息。
- 生成用户二维码。

【基本框架】

基本框架如图 7.1 到 7.6 所示。

图 7.1 主页框架图

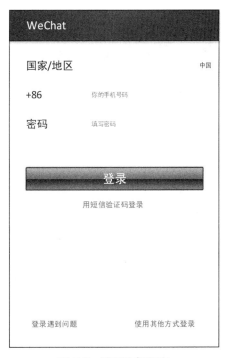

图 7.2 登录页框架图

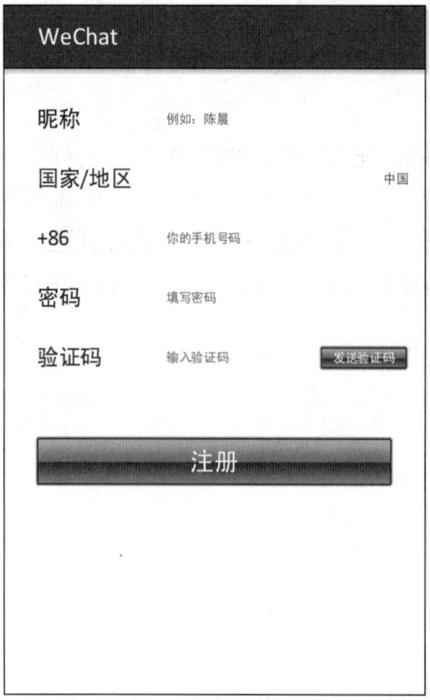

图 7.3 注册框架图

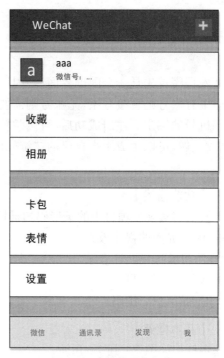

图 7.4 我的界面框架图

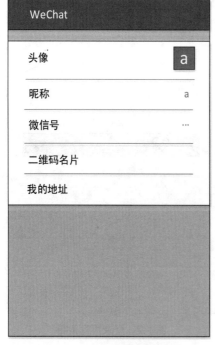

图 7.5 个人资料框架图

图 7.6 二维码框架图

通过本模块的学习，将以上的框架图转换成图 7.7 至 7.12 所示效果。

项目三 微聊

图 7.7 主页效果图

图 7.8 登录界面效果图

图 7.9 注册界面效果图

图 7.10 我的界面效果图

图 7.11　个人资料界面效果图　　　　图 7.12　二维码界面效果图

技能点一　Retrofit

在 Android 编写代码时，要想简化网络操作的工作，提高效率和正确率，这里推荐使用 Retrofit。Retrofit 是一个强大的网络库，可用于多种语言。下面将详细介绍 Retrofit。

1　Retrofit 简介

Retrofit 是 Square 公司开发的一个高质量高效率的 Http 库，它对 OkHttp 做了一层封装，是一个可以用于 Android 和 Java 的网络库，它将自己开发的底层代码和细节都封装了起来。在没有 Retrofit，OkHttp 等库的时候，可能需要自己去写 put、get、post、delete 请求。但有了这些库之后对于这些请求就只需要一行代码或者是一个注解即可。

2　Retrofit 使用介绍

使用 Retrofit 的步骤总共有七个，具体步骤如下所示：
第一步：添加 Retrofit 库的依赖；

第二步：创建接受服务器返回数据的类；
第三步：创建用于描述网络请求的接口；
第四步：创建 Retrofit 实例；
第五步：创建网络请求接口实例并配置网络请求参数；
第六步：发送网络请求（异步／同步）；
第七步：处理服务器返回的数据。
下面对每一步做一个详细的讲解。
第一步：添加 Retrofit 库的依赖

1. 在 Gradle 里面加入 Retrofit 库的依赖

由于 Retrofit 是基于 OkHttp 的，所以还需要在 build.gradle 中添加 OkHttp 库依赖。具体代码如下所示。

```
dependencies {
    compile 'com.squareup.retrofit2:retrofit:2.0.2'
    // Retrofit 库
    compile 'com.squareup.okhttp3:okhttp:3.1.2'
    // OkHttp 库
}
```

2. 添加网络权限

在 AndroidManifest.xml 中添加网络权限。具体代码如下所示。

```
<uses-permission android:name="android.permission.INTERNET"/>
```

第二步：创建接收服务器返回数据的类

在添加完网络权限之后，开始创建接收服务器数据的类：Reception.java。具体代码如下所示。

```
public class Reception {
    ...
    // 根据返回数据的格式和数据解析方式（JSON、XML 等）定义
}
```

第三步：创建用于描述网络请求的接口

Retrofit 将 Http 请求抽象成 Java 接口，采用注解描述网络请求参数和配置网络请求参数。即使用动态代理动态的方法将该接口的注解"翻译"成一个 Http 请求，最后再执行 Http 请求。这里需要注意一下，接口中的每个方法的参数都需要使用注解标注，否则会报错。

如表 7.1 所示详细介绍 Retrofit 网络请求接口的注解类型以及注解说明。

表 7.1　Retrofit 网络请求接口的注解类型

类型	注解名称	解释	作用域
网络请求方法	@GET	所有方法分别对应 Http 中的网络请求方法；都接收一个网络地址 URL（也可以不制定，通过 @Http 设置）	网络请求接口的方法
	@POST		
	@PUT		
	@DELETE		
	@PATH		
	@HEAD		
	@OPTIONS		
	@HTTP	用于替换以上 7 个注解的作用及更多功能拓展	
标记类	@FormUrlEncoded	表示请求体是一个 Form 表单	
	@Multipart	表示请求体是一个支持文件上传的 Form 表单	
	@Streaming	表示返回的数据以流的形式返回；适用于返回数据较大的场景；（如果没有使用该注解，默认把数据全部载入内存，之后获取数据也是从内存中获取）	
网络请求参数	@Headers	添加请求头	网络请求接口的方法的参数（如 Call<> getCall(*) 中的 *）
	@Header	添加不固定值得 Header	
	@Body	用于非表单请求体	
	@Field	向 Post 表单传入键值对	
	@FieldMap		
	@Part	用于表单字段，适用于有文件上传的情况	
	@PartMap		
	@Query	用于表单字段，功能同 @Field 与 @FieldMap；（区别在于它俩的数据体现在 URL 上，@Field 与 @FieldMap 的数据体现在请求体上，但生成的数据是一致的）	
	@QueryMap		
	@Path	URL 缺省值	
	@PathMap	URL 设置	

根据以上表格对各注解做一个详细的说明。

1. @GET、@POST、@PUT、@DELETE、@HEAD。

这几个注解分别对应 HTTP 中的网络请求方式。示例如下所示。

```
public interface GetRequest_Interface {
    @GET("openapi.do?keyfrom=Yanzhikai&key=2032414398&type=data&doc-type=json&version=1.1&q=car")
    Call<Translation> getCall();
```

```
// @GET 注解的作用：采用 Get 方法发送网络请求
// getCall() = 接收网络请求数据的方法
// 其中返回类型为 Call<*>,* 是接收数据的类（即上面定义的 Translation 类）
// 如果想直接获得 Responsebody 中的内容,可以定义网络请求返回值为
   Call<ResponseBody>
}
```

此处说明 URL 的组成,Retrofit 把网络请求的 URL 分成了两部分设置,具体如下所示。

```
// 第 1 部分：在网络请求接口的注解设置
   @GET("openapi.do?keyfrom=Yanzhikai&key=2032414398&type=data&doctype=-
json&version=1.1&q=car")
   Call<Translation>  getCall();
// 第 2 部分：在创建 Retrofit 实例时通过 .baseUrl() 设置
Retrofit retrofit = new Retrofit.Builder()
        .baseUrl("http://fanyi.youdao.com/") // 设置网络请求的 Url 地址
        .addConverterFactory(GsonConverterFactory.create()) // 设置数据解析器
        .build();
// 从上面看出：一个请求的 URL 可以通过替换块和请求方法的参数来进行动
// 态的 URL 更新。
// 替换块是由被 {} 包裹起来的字符串构成
// 即：Retrofit 支持动态改变网络请求根目录
```

注意：在网络请求当中,完整的 Url= 在创建 Retrofit 实例时通过 .baseUrl() 设置 + 网络请求接口的注解设置（以下称"path"）。具体整合的规则如表 7.2 所示。

表 7.2　整合规则

类型	具体使用
path= 完整的 Url	Url="http://host:post/aa/apath"，其中 path="http://host:post/aa/apath"，baseUrl= 不设置（如果接口里的 Url 是一个完整的网址,那么在创建 Retrofit 实例时可以不设置 URL）
path= 绝对路径	Url="http://host:post/apath"，其中 path="/apath"，baseUrl="http://host:post/a/b/"
path= 相对路径 baseUrl= 目录形式	Url="http://host:post/a/b/apath"，其中 path="apath"，baseUrl="http://host:post/a/b/"
path= 相对路径 baseUrl= 文件形式	Url="http://host:post/a/ apath"，其中 path="apath"，baseUrl="http://host:post/a/b/"

通常情况建议使用第三种方式来进行配置,并且使用同一种路径形式。

2. @HTTP

可以替换 @GET、@POST、@PUT、@DELETE、@HEAD 注解的作用及更多功能拓展，通过属性 method、path、hasBody 来进行设置。示例如下所示。

```
public interface GetRequest_Interface {
    /**
     * method:网络请求的方法(区分大小写)
     * path:网络请求地址路径
     * hasBody:是否有请求体
     */
    @HTTP(method = "GET", path = "blog/{id}", hasBody = false)
    Call<ResponseBody> getCall(@Path("id") int id);
    // {id} 表示是一个变量
    // method 的值 retrofit 不会做处理，所以要自行保证准确
}
```

3. @FormUrlEncoded

表示发送 form-encoded 的数据，每个键值对需要用 @Filed 来注解键名，随后的对象需要提供值。

4. @Multipart

表示发送 form-encoded 的数据（适用于有文件上传的场景），每个键值对需要用 @Part 来注解键名，随后的对象需要提供值。具体使用如下。

```
public interface GetRequest_Interface {
    /*
表明是一个表单格式的请求（Content-Type:application/x-www-form-urlencoded）
<code>Field("username")</code> 表示将后面的
<code>String name</code> 中 name 的取值作 * 为 username 的值
    */
    @POST("/form")
    @FormUrlEncoded
    Call<ResponseBody> testFormUrlEncoded1(@Field("username") String name,
        @Field("age") int age);
        /*
        {@link Part} 后面支持三种类型，{@link RequestBody}、{@link okhttp3.MultipartBody.Part} 、* 任意类型除 {@link okhttp3.MultipartBody.Part} 以外，其它类型都必须带上表单字段 ({@link *okhttp3.MultipartBody.Part} 中已经包含了表单字段的信息)
        */
    @POST("/form")
```

```
        @Multipart
        Call<ResponseBody> testFileUpload1(@Part("name")
    RequestBody name, @Part("age") RequestBody age,
        @Part MultipartBody.Part file);
}
    // 具体使用
    GetRequest_Interface service = retrofit.create(GetRequest_Interface.class);
        // @FormUrlEncoded
    Call<ResponseBody> call1 = service.testFormUrlEncoded1("Carson", 24);
        // @Multipart
        RequestBody name = RequestBody.create(textType, "Carson");
        RequestBody age = RequestBody.create(textType, "24");
        MultipartBody.Part filePart = MultipartBody.Part.createFormData(
        "file", "test.txt", file);
        Call<ResponseBody> call3 = service.testFileUpload1(name, age, filePart);
```

5. @Header&@Headers

用于添加请求头 & 添加不固定的请求头，具体使用如下所示。

```
    // @Header
    @GET("user")
    Call<User> getUser(@Header("Authorization") String authorization)
    // @Headers
    @Headers("Authorization: authorization")
    @GET("user")
    Call<User> getUser()
    // 以上的效果是一致的。
    // 区别在于使用场景和使用方式
    // 1. 使用场景：@Header 用于添加不固定的请求头，
    //@Headers 用于添加固定的请求头
    // 2. 使用方式：@Header 作用于方法的参数；@Headers 作用于方法
```

6. @Body

以 Post 方式传递自定义数据类型给服务器，要注意如果提交的是一个 Map，它的作用相当于 @Field。不过 Map 要经过 FormBody.Builder 类处理成为符合 OkHttp 格式的表单，示例如下所示。

```
    FormBody.Builder builder = new FormBody.Builder();
    builder.add("key","value");
```

7. @Field&FieldMap

在发送 Post 请求时提交请求的表单字段，使用时与 @FormUrlEncoded 注解配合使用。示例如下所示。

```java
public interface GetRequest_Interface {
    /**
     * 表明是一个表单格式的请求（Content-Type:
     *application/x-www-form-urlencoded）
     *<code>Field("username")</code> 表示将后面的 <code>String name
     *</code> 中 name 的取值
     * 作为 username 的值
     */
    @POST("/form")
    @FormUrlEncoded
    Call<ResponseBody> testFormUrlEncoded1(@Field("username")
    String name, @Field("age") int age);
    /**
     * Map 的 key 作为表单的键
     */
    @POST("/form")
    @FormUrlEncoded
    Call<ResponseBody> testFormUrlEncoded2(@FieldMap Map<String, Object> map);}
    // 具体使用
    // @Field
    Call<ResponseBody> call1 = service.testFormUrlEncoded1("Carson", 24);
    // @FieldMap
    // 实现的效果与上面相同，但要传入 Map
    Map<String, Object> map = new HashMap<>();
    map.put("username", "Carson");
    map.put("age", 24);
    Call<ResponseBody> call2 = service.testFormUrlEncoded2(map);
```

8. @Part&&PartMap

在发送 Post 请求时提交请求的表单字段，与 @Field 的区别是功能相同，但携带的参数类型更加丰富，包括数据流，所以适用于有文件上传的场景，使用时与 @Multipart 注解配合使用。

9. Query 和 @QueryMap

用于 @GET 方法的查询参数（Query=Url 中 '?' 后面的 key-value），如：url=http://www.println.net/?cate=android，其中，Query=cate。在配置时只需要在接口方法中增加一个参数即可。

示例如下所示。

```
@GET("/")
    Call<String> cate(@Query("cate") String cate);
// 其使用方式同 @Field 与 @FieldMap，这里不作过多描述
```

10. @Path

URL 地址的缺省值，示例如下所示。

```
public interface GetRequest_Interface {
        @GET("users/{user}/repos")
        Call<ResponseBody> getBlog(@Path("user") String user );
        // 访问的 API 是：https://api.github.com/users/{user}/repos
        // 在发起请求时，{user} 会被替换为方法的第一个参数 user
        //（被 @Path 注解作用）
    }
```

11. @Url

直接传入一个请求的 URL 变量用于 URL 设置。示例如下所示。

```
public interface GetRequest_Interface {
    @GET
    Call<ResponseBody> testUrlAndQuery(@Url String url, @Query("showAll") boolean showAll);
// 当有 URL 注解时，@GET 传入的 URL 就可以省略
// 当 GET、POST...HTTP 等方法中没有设置 Url 时，则必须使用 {@link Url} 提供
    }
```

第四步：创建 Retrofit 实例

创建接口之后接着要创建 Retrofit 实例，示例如下所示。

```
Retrofit retrofit = new Retrofit.Builder()
        .baseUrl("http://fanyi.youdao.com/") // 设置网络请求的 Url 地址
        .addConverterFactory(GsonConverterFactory.create()) // 设置数据解析器
        .addCallAdapterFactory(RxJavaCallAdapterFactory.create())// 支持 RxJava 平台
        .build();
```

根据示例可知，创建 Retrofit 时要设置数据解析器。Retrofit 支持多种数据解析方式，但在使用时要在 Gradle 中添加依赖。如表 7.3 所示是各种数据解析器以及对应的 Gradle 依赖。

表 7.3　各种数据解析器以及对应的 Gradle 依赖

数据解析器	Gradle 依赖
Gson	com.squareup.retrofit2:converter-gson:2.0.2
Jackson	com.squareup.retrofit2:converter-jackson:2.0.2
Simple XML	com.squareup.retrofit2:converter-simplexml:2.0.2
Protobuf	com.squareup.retrofit2:converter-protobuf:2.0.2
Moshi	com.squareup.retrofit2:converter-moshi:2.0.2
Wire	com.squareup.retrofit2:converter-wire:2.0.2
Scalars	com.squareup.retrofit2:converter-scalars:2.0.2

同时 Retrofit 支持多种网络请求适配方式，有：guava、Java8 和 RxJava。在使用时如果是 Android 默认的 CallAdapter，不需要添加网络请求适配器的依赖，否则需要按照需求进行添加 Retrofit 提供的 CallAdapter。如表 7.4 所示是使用时需要在 Gradle 中添加的依赖。

表 7.4　使用时需要在 Gradle 中添加的依赖

网络请求适配器	Gradle 依赖
guava	com.squareup.retrofit2:adapter-guava:2.0.2
Java8	com.squareup.retrofit2:adapter-java8:2.0.2
Rxjava	com.squareup.retrofit2:adapter-rxjava:2.0.2

第五步：创建网络请求接口实例

示例如下所示。

```
// 创建网络请求接口的实例
GetRequest_Interface request = retrofit.create(GetRequest_Interface.class);
// 对发送请求进行封装
Call<Reception> call = request.getCall();
```

第六步：发送网络请求（异步/同步）

这一步主要是封装了数据转换、线程切换的操作。示例如下所示。

```
// 发送网络请求 ( 异步 )
call.enqueue(new Callback<Translation>() {
    // 请求成功时回调
    @Override
    public void onResponse(Call<Translation> call,
    Response<Translation> response) {
```

```
            // 请求处理，输出结果
            response.body().show();
        }
        // 请求失败时候的回调
        @Override
        public void onFailure(Call<Translation> call, Throwable throwable) {
            System.out.println(" 连接失败 ");
        }
    });
// 发送网络请求（同步）
Response<Reception> response = call.execute();
```

第七步：处理返回数据

最后要处理服务器返回的数据，是通过 response 类的 body() 对返回的数据进行处理。示例如下所示。

```
        // 发送网络请求（异步）
        call.enqueue(new Callback<Translation>() {
            // 请求成功时回调
            @Override
            public void onResponse(Call<Translation> call,
                Response<Translation> response) {
                // 对返回数据进行处理
                response.body().show();
            }
            // 请求失败时候的回调
            @Override
            public void onFailure(Call<Translation> call, Throwable throwable) {
                System.out.println(" 连接失败 ");
            }
        });
// 发送网络请求（同步）
Response<Reception> response = call.execute();
// 对返回数据进行处理
response.body().show();
```

技能点二 MVP

在刚开始学做项目时，都会将所有的代码（如数据的访问和处理、数据的展示、用户的输入）写在一起。代码和思维都呈现出一种线性的形式，这时要思考如何把组合在一起的代码拆分，如何把拆分的代码再组合。想要达到这个目标，就需要学习一些架构，例如 MVP、MVC 等。

1 MVP 简介

MVP 是 Model-View-Presenter 的简称，分别表示数据层、视图层、发布层。Model 是用来存放数据的处理（比如网络请求，缓存等），View 是负责 UI 界面的实现，Presenter 则负责处理业务逻辑代码，处理 Model 数据，然后将处理完的数据分发到 View 层。由于在创建项目时，代码都是在 Activity 中，导致 Activity 代码太多，不利于后期修改维护。MVP 架构则具有良好的模块分工，可以大大简化对代码的理解程度。如图 7.13 所示是 MVP 的结构图。

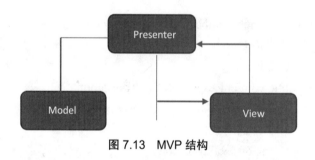

图 7.13 MVP 结构

2 MVP 的工作模式

MVP 的主要特点就是把 Activity 里的许多逻辑抽离到 View 和 Presenter 接口中去，并由具体的实现类来完成。在 MVP 模式中 Activity 的功能就是响应生命周期和显示界面，其他的工作都在 Presenter 层中进行完成，Presenter 是 Model 层和 View 层的桥梁。它具体的步骤如图 7.14 所示：

创建 Presenter 接口，把所有业务逻辑的接口都放在这里，并创建它的实现 PresenterCompl。

创建 IView 接口，把所有视图逻辑的接口都放在这里，其实现类是当前的 Activity/Fragment。

由 UML 图可以看出，Activity 里包含了一个 IPresenter，而 PresenterCompl 里又包含了一个 IView，并且依赖 Model。Activity 里只保留对 IPresenter 的调用，其它工作全部留到 PresenterCompl 中实现。

Model 并不是必须有的，但是一定会有 View 和 Presenter。

项目三 微聊   197

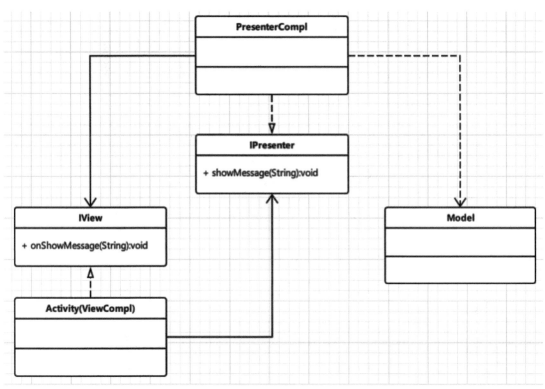

图 7.14 UML 图

3 MVP 的使用

1. 在 MVP 的实践中,将之前的一个 Activity 转换成 3 个类来实现。项目中结构图如图 7.15 所示。

图 7.15 MVP 结构图

2. 结构中代码如下：

Modle 负责数据和业务逻辑。具体代码如下所示。

```java
public interface User{
    void login(String username, String password, Callback callback);
}
```

View 负责展示数据。具体代码如下所示。

```java
public class UserActivity extends Activity implements UserView {
    private UserContract.Presenter mPresenter;
    private TextView mUsernameTextView;
    private TextView mPasswordTextView;
    private Button mLoginBtn;
    @Override
    public void onCreate(Bundle savedInstanceState) {
        super.onCreate(savedInstanceState);
        setContentView(R.layout.login);
        mPresenter = new AddressListPresenter(this, new UserModelImpl());
        mUsernameTextView = (TextView) findViewById(R.id.username);
        mPasswordTextView = (TextView) findViewById(R.id.password);
        mLoginBtn = (Button) findViewById(R.id.login_btn);
        mLoginBtn.setOnClickListener(new View.OnClickListener() {
            public void onClick(View v)  {
                String username = mUsernameTextView.getText().toString();
                String password = mPasswordTextView.getText().toString();
                // View 将用户的点击事件直接路由给 Presenter 区处理
                mPresenter.login(username, password);
            } });  }
    @Override
    public void showLoginSuccessMsg(User loginedUser) {
    //Presenter 在处理完毕后，会通知 View 更新 UI 来通知用户数据操作的结果
        Toast.makeToast(getApplicationContext(), "Login Success",
            Toast.LENGTH_SHORT).show();
}
    @Override
    public void showLoginFailMsg(String errorMsg) {
    // Presenter 在处理完毕后，会通知 View 更新 UI 来通知用户数据操作的结果
    Toast.makeToast(getApplicationContext(), "Login Fail",
            Toast.LENGTH_SHORT).show();
}
```

```java
@Override
protected void onResume() {
    super.onResume();
    mPresenter.subscribe();
}
@Override
protected void onPause() {
    super.onPause();
    mPresenter.unSubscribe();
} }
```

Presenter 负责做 View 和 Model 的连接。具体代码如下所示。

```java
public class UserPresenterImpl implements UserPresenter {
    private UserView mUserView;
    private UserModel mUserModel;
    public UserPresenterImpl(UserView view, UserModel model) {
        mUserView = view;
        mUserModel = model;
    }
    public void login(String username, String password) {
        // Presenter 处理 View 路由过来的用户操作,
        // 将其转换成相对的命令,传递给 Model 来做数据操作
        mUserModel.login(username, password, new Callback(){
            public void onSuccess(User user) {
                // Model 层对数据操作后,将结果返回给 Presenter,
                // 再由 Presenter 来通知 View 去更新 UI 来通知
                // 用户数据操作的结果
                mView.showLoginSuccessMsg(user);
            }
            public void onFail(String errorMsg) {
                mView.showLoginFailMsg(user);
            } }); }
```

4. MVP 优势

MVP 非常适合大型的 APP 开发,越复杂它的优势越明显。其中的优势有:

一、分离了视图逻辑和业务逻辑,降低了耦合。

二、Activity 只处理生命周期的任务,代码变得更加简洁。

三、视图逻辑和业务逻辑分别抽象到 View 和 Presenter 的接口中去,提高代码的可阅

读性。

四、Presenter 被抽象成接口，可以有多种具体的实现，所以方便进行单元测试。

五、把业务逻辑抽到 Presenter 中去，避免后台线程引用 Activity 导致 Activity 的资源无法被系统回收从而引起内存泄露和 OOM。

拓展：通过以上内容的学习，已经了解了 Android MVP 框架模式。谷歌 Android 开发团队鼓励开发者利用 MVP 框架模式开发项目，我们平时写代码也或多或少的在使用 MVP 框架模式开发项目。在日常的 Android 项目开发过程中有许多框架供我们选择，扫描右侧二维码即可了解详情。

通过以上技能点的学习，使用 Retrofit、MVP 框架编写代码，并且实现微聊 App 的系统及个人功能。

1. 系统

第一步：在运行项目时，首先看到的是闪屏界面，编写闪屏。具体代码如 CORE0701 所示。

代码 CORE0701　闪屏功能的实现

```java
public class SplashActivity extends BaseActivity {
//Activity 可能用到的控件
@Bind(R.id.rlButton)
RelativeLayout mRlButton;
@Bind(R.id.btnLogin)
Button mBtnLogin;
@Bind(R.id.btnRegister)
Button mBtnRegister;
@Override
public void init() {
// 动态获取不安全的权限
    PermissionGen.with(this)
        .addRequestCode(100)
        .permissions(
            // 电话通讯录
            Manifest.permission.GET_ACCOUNTS,
            Manifest.permission.READ_PHONE_STATE,
            // 位置
            Manifest.permission.ACCESS_FINE_LOCATION,
            Manifest.permission.ACCESS_COARSE_LOCATION,
```

```java
                        Manifest.permission.ACCESS_FINE_LOCATION,
                        // 相机、麦克风
                        Manifest.permission.RECORD_AUDIO,
                        Manifest.permission.WAKE_LOCK,
                        Manifest.permission.CAMERA,
                        // 存储空间
                        Manifest.permission.WRITE_EXTERNAL_STORAGE,
                        Manifest.permission.WRITE_SETTINGS
                )
                .request();
        if (!TextUtils.isEmpty(UserCache.getToken())) {
            Intent intent = new Intent(this, MainActivity.class);
            intent.setFlags(Intent.FLAG_ACTIVITY_CLEAR_TASK);
            jumpToActivity(intent);
            finish();
        }
    }
    @Override
    public void initView() {
        StatusBarUtil.setColor(this, UIUtils.getColor(R.color.black));
        AlphaAnimation alphaAnimation = new AlphaAnimation(0, 1); // 闪屏动画
        alphaAnimation.setDuration(1000);
        mRlButton.startAnimation(alphaAnimation);
    }
    @Override
    public void initListener() {
        mBtnLogin.setOnClickListener(v -> {
            jumpToActivity(LoginActivity.class);    // 跳转至登录界面
            finish();
        });
        mBtnRegister.setOnClickListener(v -> {
            jumpToActivity(RegisterActivity.class);  // 跳转至注册界面
            finish();
        });
    }
    @Override
    protected BasePresenter createPresenter() {
        return null;
    }
}
```

```
@Override
protected int provideContentViewId() {        // 对应的布局文件
    return R.layout.activity_splash;
}}
```

完成上述代码,实现效果如图 7.16 所示。

图 7.16　主界面效果图

第二步:登录界面的实现,如果已经有账号,则直接登录,如果无账号则进行注册。具体代码如 CORE0702 所示。

代码 CORE0702　　登录界面的实现
/***@ 描述 登录界面 */ public class LoginActivity extends BaseActivity<ILoginAtView, LoginAtPresenter> implements ILoginAtView {　// 以下是 Activity 中可能用到的控件 　　@Bind(R.id.ibAddMenu) 　　ImageButton mIbAddMenu; 　　@Bind(R.id.etPhone) 　　EditText mEtPhone; 　　@Bind(R.id.vLinePhone) 　　View mVLinePhone; 　　@Bind(R.id.etPwd)

```java
EditText mEtPwd;
@Bind(R.id.vLinePwd)
View mVLinePwd;
@Bind(R.id.tvProblems)
TextView mTvProblems;
@Bind(R.id.btnLogin)
Button mBtnLogin;
@Bind(R.id.tvOtherLogin)
TextView mTvOtherLogin;
// 文本监听接口,修改当前文本的内容
TextWatcher watcher = new TextWatcher() {
@Override
// 文本没有被改变,但是即将被改变
public void beforeTextChanged(CharSequence s, int start, int count, int after) {

    }
@Override
// 文本被改变时,改变结果已将显示
public void onTextChanged(CharSequence s, int start, int before, int count) {
        mBtnLogin.setEnabled(canLogin());
    }
    @Override
    // 文本已经被修改完成
    public void afterTextChanged(Editable s) {
    }  };
@Override
public void initView() {
    mIbAddMenu.setVisibility(View.GONE);
}
@Override
public void initListener() {
    // 将文本加入监听
    mEtPwd.addTextChangedListener(watcher);
    mEtPhone.addTextChangedListener(watcher);
    // 焦点事件
    mEtPwd.setOnFocusChangeListener((v, hasFocus) -> {
        // 判断焦点是否存在
        if (hasFocus) {
            mVLinePwd.setBackgroundColor(UIUtils.getColor(R.color.green0));
```

```java
            } else {
                mVLinePwd.setBackgroundColor(UIUtils.getColor(R.color.line));
            }
        });
        mEtPhone.setOnFocusChangeListener((v, hasFocus) -> {
            if (hasFocus) {
                mVLinePhone.setBackgroundColor(UIUtils.getColor(R.color.green0));
            } else {
                mVLinePhone.setBackgroundColor(UIUtils.getColor(R.color.line));
            }
        });
        // 点击监听
        mBtnLogin.setOnClickListener(v -> mPresenter.login());
}
// 是否输入了用户名和密码
private boolean canLogin() {
    int pwdLength = mEtPwd.getText().toString().trim().length();
    int phoneLength = mEtPhone.getText().toString().trim().length();
    if (pwdLength > 0 && phoneLength > 0) {
        return true;      }
    return false;    } }
```

完成上述代码，实现效果如图 7.17 所示。

图 7.17 登录界面效果图

第三步:通过网络请求验证用户名密码是否正确。具体代码如 CORE0703 所示。

代码 CORE0703　验证登录密码

```java
public class LoginAtPresenter extends BasePresenter<ILoginAtView> {
    // 验证登录的初始化方法
    public LoginAtPresenter(BaseActivity context) {
        super(context);
    }
    public void login() {
        String phone = getView().getEtPhone().getText().toString().trim();
        String pwd = getView().getEtPwd().getText().toString().trim();
        // 判断用户名和密码
        if (TextUtils.isEmpty(phone)) {
            UIUtils.showToast(UIUtils.getString(R.string.phone_not_empty));
            return;
        }
        if (TextUtils.isEmpty(pwd)) {
            UIUtils.showToast(UIUtils.getString(R.string.password_not_empty));
            return;
        }
        mContext.showWaitingDialog(UIUtils.getString(R.string.please_wait));
        // 通过网络请求,查询用户名和密码
        ApiRetrofit.getInstance().login(AppConst.REGION, phone, pwd)
                .subscribeOn(Schedulers.io())
                .observeOn(AndroidSchedulers.mainThread())
                .subscribe(loginResponse -> {
                    int code = loginResponse.getCode();
                    mContext.hideWaitingDialog();
                    if (code == 200) {
                        UserCache.save(loginResponse.getResult().getId(), phone,
                        loginResponse.getResult().getToken());
                        mContext.jumpToActivityAndClearTask(MainActivity.class);
                        mContext.finish();
                    } else {
                        // 登录失败的反应
                        loginError(new ServerException(UIUtils.getString(R.string.login_error) + code));
                    }
                }, this::loginError);    }
```

```java
private void loginError(Throwable throwable) {
    // 失败后的事件处理
        LogUtils.e(throwable.getLocalizedMessage());
        UIUtils.showToast(throwable.getLocalizedMessage());
        mContext.hideWaitingDialog();
    }}
```

第四步：如果没有账号进行登录，则需要进行注册。具体代码如 CORE0704 所示。

代码 CORE0704　注册界面的实现

```java
public class RegisterActivity extends BaseActivity<IRegisterAtView, RegisterAtPresenter> implements IRegisterAtView {
    //Activity 中可能用到的控件
    @Bind(R.id.etNick)
    EditText mEtNick;
    @Bind(R.id.vLineNick)
    View mVLineNick;
    @Bind(R.id.etPhone)
    EditText mEtPhone;
    @Bind(R.id.vLinePhone)
    View mVLinePhone;
    @Bind(R.id.etPwd)
    EditText mEtPwd;
    @Bind(R.id.ivSeePwd)
    ImageView mIvSeePwd;
    @Bind(R.id.vLinePwd)
    View mVLinePwd;
    @Bind(R.id.etVerifyCode)
    EditText mEtVerifyCode;
    @Bind(R.id.btnSendCode)
    Button mBtnSendCode;
    @Bind(R.id.vLineVertifyCode)
    View mVLineVertifyCode;
    @Bind(R.id.btnRegister)
    Button mBtnRegister;
    // 文本监听接口
    TextWatcher watcher = new TextWatcher() {
        @Override
```

```java
        public void beforeTextChanged(CharSequence s, int start,
        int count, int after) {
        }
        @Override
        public void onTextChanged(CharSequence s, int start,
        int before, int count) {
            mBtnRegister.setEnabled(canRegister());
        }
        @Override
        public void afterTextChanged(Editable s) {
        }   };
@Override
public void initListener() {
    mEtNick.addTextChangedListener(watcher);
    mEtPwd.addTextChangedListener(watcher);
    mEtPhone.addTextChangedListener(watcher);
    mEtVerifyCode.addTextChangedListener(watcher);
    mEtNick.setOnFocusChangeListener((v, hasFocus) -> {
        if (hasFocus) {
            mVLineNick.setBackgroundColor(UIUtils.getColor(R.color.green0));
        } else {
            mVLineNick.setBackgroundColor(UIUtils.getColor(R.color.line));
        }       });
    mEtPwd.setOnFocusChangeListener((v, hasFocus) -> {
        if (hasFocus) {
            mVLinePwd.setBackgroundColor(UIUtils.getColor(R.color.green0));
        } else {
            mVLinePwd.setBackgroundColor(UIUtils.getColor(R.color.line));
        }       });
    mEtPhone.setOnFocusChangeListener((v, hasFocus) -> {
        if (hasFocus) {
            mVLinePhone.setBackgroundColor(
                UIUtils.getColor(R.color.green0));
        } else {
            mVLinePhone.setBackgroundColor(
                UIUtils.getColor(R.color.line));
        }       });
    mEtVerifyCode.setOnFocusChangeListener((v, hasFocus) -> {
```

```java
            if (hasFocus) {
                mVLineVertifyCode.setBackgroundColor(
                    UIUtils.getColor(R.color.green0));
            } else {
                mVLineVertifyCode.setBackgroundColor(
                    UIUtils.getColor(R.color.line));
            }
        });
        mIvSeePwd.setOnClickListener(v -> {
            // 密码的隐藏和显示
            if (mEtPwd.getTransformationMethod() ==
                HideReturnsTransformationMethod.getInstance()) {
                mEtPwd.setTransformationMethod(
                    PasswordTransformationMethod.getInstance());
            } else {
                mEtPwd.setTransformationMethod(
                    HideReturnsTransformationMethod.getInstance());
            }
            mEtPwd.setSelection(mEtPwd.getText().toString().trim().length());
        });
        mBtnSendCode.setOnClickListener(v -> {
            // 是否存在焦点,并执行接收 OnClick 事件
            if (mBtnSendCode.isEnabled()) {
                mPresenter.sendCode();
            }
        });
        mBtnRegister.setOnClickListener(v -> {
            mPresenter.register();
        });    }
@Override
protected void onDestroy() {
    super.onDestroy();
    mPresenter.unsubscribe();
}
private boolean canRegister() {        // 判断是否能进行登录
    int nickNameLength = mEtNick.getText().toString().trim().length();
    int pwdLength = mEtPwd.getText().toString().trim().length();
    int phoneLength = mEtPhone.getText().toString().trim().length();
    int codeLength = mEtVerifyCode.getText().toString().trim().length();
```

```
    if (nickNameLength > 0 && pwdLength > 0 && phoneLength > 0
        && codeLength > 0) {
        return true;    }
    return false;    }
```

完成上述代码,实现效果如图 7.18 所示。

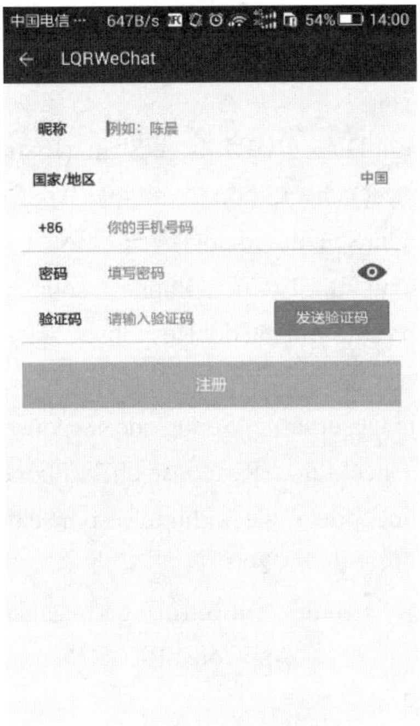

图 7.18 注册界面效果图

第五步:通过网络请求,将注册的用户名密码进行保存。具体代码如 CORE0705 所示。

代码 CORE0705   注册信息的保存

```
public class RegisterAtPresenter extends BasePresenter<IRegisterAtView> {
int time = 0;
private Timer mTimer;
private Subscription mSubscription;
// 初始化方法
public RegisterAtPresenter(BaseActivity context) {
    super(context);
}
public void sendCode() {
    String phone = getView().getEtPhone().getText().toString().trim();
    // 判断手机号是否正确
```

```java
if (TextUtils.isEmpty(phone)) {
    UIUtils.showToast(UIUtils.getString(R.string.phone_not_empty));
    return;
}
if (!RegularUtils.isMobile(phone)) {
    UIUtils.showToast(UIUtils.getString(R.string.phone_format_error));
    return;
}
// 等待短信接收倒计时
mContext.showWaitingDialog(UIUtils.getString(R.string.please_wait));
ApiRetrofit.getInstance().checkPhoneAvailable(AppConst.REGION, phone)
        .subscribeOn(Schedulers.io())
        .flatMap(new Func1<CheckPhoneResponse,
            Observable<SendCodeResponse>>() {
            @Override
            public Observable<SendCodeResponse> call(
                CheckPhoneResponse checkPhoneResponse) {
                int code = checkPhoneResponse.getCode();
                if (code == 200) {
                    return ApiRetrofit.getInstance().sendCode(
                        AppConst.REGION, phone);
                } else {
                    return Observable.error(new ServerException(
                        UIUtils.getString(R.string.phone_not_available)));
                }
            } })
        // 接收验证码
        .observeOn(AndroidSchedulers.mainThread())
        .subscribe(sendCodeResponse -> {
            mContext.hideWaitingDialog();
            int code = sendCodeResponse.getCode();
            if (code == 200) {
                changeSendCodeBtn();
            } else {
                sendCodeError(new ServerException(
                    UIUtils.getString(R.string.send_code_error)));
            }
        }, this::sendCodeError);    }
```

```java
private void changeSendCodeBtn() {
    // 开始 1 分钟倒计时
    // 每一秒执行一次 Task
    mSubscription = Observable.create((Observable.OnSubscribe<Integer>)
    subscriber -> {
        time = 60;
        TimerTask mTask = new TimerTask() {
            @Override
            public void run() {
                subscriber.onNext(--time);
            }                      };
        mTimer = new Timer();
        mTimer.schedule(mTask, 0, 1000);
        // 每一秒执行一次 Task
    }).subscribeOn(Schedulers.io())
        .observeOn(AndroidSchedulers.mainThread())
        .subscribe(time -> {
            if (getView().getBtnSendCode() != null) {
                if (time >= 0) {
                    getView().getBtnSendCode().setEnabled(false);
                    getView().getBtnSendCode().setText(time + "");
                } else {
                    getView().getBtnSendCode().setEnabled(true);
                    getView().getBtnSendCode().setText(
                        UIUtils.getString(R.string.send_code_btn_normal_tip));
                }  } else {
                mTimer.cancel();}
        }, throwable -> LogUtils.sf(throwable.getLocalizedMessage()));    }
ApiRetrofit.getInstance().verifyCode(AppConst.REGION, phone, code)
            // 接收的验证码是否正确
            .flatMap(new Func1<VerifyCodeResponse,
                    Observable<RegisterResponse>>() {
                @Override
                public Observable<RegisterResponse>
                    call(VerifyCodeResponse verifyCodeResponse) {
                    int code = verifyCodeResponse.getCode();
                    if (code == 200) {
```

```java
                            return ApiRetrofit.getInstance().register(nickName,
                                password, verifyCodeResponse.getResult().
                                getVerification_token());
                        } else {
                            return Observable.error(new
                                ServerException(UIUtils.getString(
                                R.string.vertify_code_error) + code));
                        }
                    }
                })
                // 手机号与验证码是否一致
                .flatMap(new Func1<RegisterResponse, Observable<LoginResponse>>() {
                    @Override
                    public Observable<LoginResponse> call(
                        RegisterResponse registerResponse) {
                        int code = registerResponse.getCode();
                        if (code == 200) {
                            return ApiRetrofit.getInstance().login(
                                AppConst.REGION, phone, password);
                        } else {
                            return Observable.error(new
                                ServerException(UIUtils.getString(
                                R.string.register_error) + code));
                        }
                    }
                })
                .subscribeOn(Schedulers.io())
                .observeOn(AndroidSchedulers.mainThread())
                .subscribe(loginResponse -> {
                    // 判断正确后登陆成功,错误后登陆失败
                    int responseCode = loginResponse.getCode();
                    if (responseCode == 200) {
                    UserCache.save(loginResponse.getResult().getId(), phone,
                    loginResponse.getResult().getToken());
                    mContext.jumpToActivityAndClearTask(MainActivity.class);
                        mContext.finish();
                    } else {
                        UIUtils.showToast(UIUtils.getString(R.string.login_error));
                        mContext.jumpToActivity(LoginActivity.class);
                    }
                }, this::registerError);      }
```

第六步:退出当前账号与关闭应用。具体代码如 CORE0706 所示。

代码 CORE0706　退出处理

```java
public class SettingActivity extends BaseActivity {
    private View mExitView;
    @Bind(R.id.oivExit)
    OptionItemView mOivExit;
    private CustomDialog mExitDialog;
    @Override
    public void initListener() {
        mOivExit.setOnClickListener(v -> {
            // 退出的点击事件
            if (mExitView == null) {
                mExitView = View.inflate(this, R.layout.dialog_exit, null);
                // 退出当前账号和退出 APP 的弹窗
                mExitDialog = new CustomDialog(this, mExitView, R.style.MyDialog);
                // 退出当前账号
                mExitView.findViewById(R.id.tvExitAccount).setOnClickListener(v1 -> {
                    RongIMClient.getInstance().logout();
                    UserCache.clear();
                    mExitDialog.dismiss();
                    MyApp.exit();
                    jumpToActivityAndClearTask(LoginActivity.class);
                });
                // 退出 APP
                mExitView.findViewById(R.id.tvExitApp).setOnClickListener(v1 -> {
                    RongIMClient.getInstance().disconnect();
                    mExitDialog.dismiss();
                    MyApp.exit();
                });
            }
            mExitDialog.show();
        });
    }
    @Override
    protected BasePresenter createPresenter() {
        return null;
    }
    @Override
    protected int provideContentViewId() {    // 对应界面的布局
        return R.layout.activity_setting;
    }
}
```

2. 个人

系统登录成功后,需要进入个人信息模块,并完成个人信息的完善。

第一步:查看头像功能实现。具体代码如 CORE0707 所示。

代码 CORE0707　查看头像

```java
public class ShowBigImageActivity extends BaseActivity {
    private String mUrl;
    //Activity 中要用到的控件
    @Bind(R.id.ibToolbarMore)
    ImageButton mIbToolbarMore;
    @Bind(R.id.pv)
    PhotoView mPv;
    @Bind(R.id.pb)
    ProgressBar mPb;
    private FrameLayout mView;
    private PopupWindow mPopupWindow;
    @Override
    public void init() {
        mUrl = getIntent().getStringExtra("url");    // 获取拍照成功后的拍照路径
    }
    @Override
    public void initView() {
    // 修改标题
        setToolbarTitle(UIUtils.getString(R.string.header_pic));
        mIbToolbarMore.setVisibility(View.VISIBLE);
        if (TextUtils.isEmpty(mUrl)) {
            finish();
            return;
        }
        mPv.enable();// 启用图片缩放功能
        Glide.with(this).load(Uri.parse(mUrl)).placeholder(R.mipmap.default_image).centerCrop().into(mPv);
    }
    @Override
    public void initListener() {
        mIbToolbarMore.setOnClickListener(v -> showPopupMenu());    // 标签监听器
    }
    @Override
```

```java
    protected BasePresenter createPresenter() {
        return null;
    }
    @Override
    protected int provideContentViewId() {
        return R.layout.activity_show_big_image;   // 当前界面的布局文件
    }
    private void showPopupMenu() {
        if (mView == null) {
            // 点击头像,进入头像页
            mView = new FrameLayout(this);
            mView.setLayoutParams(new ViewGroup.LayoutParams(
                    ViewGroup.LayoutParams.MATCH_PARENT,
                    ViewGroup.LayoutParams.MATCH_PARENT));
            mView.setBackgroundColor(UIUtils.getColor(R.color.white));
            // 点击菜单设置按钮,会提示"保存到手机"
            TextView tv = new TextView(this);
            FrameLayout.LayoutParams params = new
                    FrameLayout.LayoutParams(
                    FrameLayout.LayoutParams.MATCH_PARENT, UIUtils.dip2Px(45));
            tv.setLayoutParams(params);
            tv.setGravity(Gravity.LEFT | Gravity.CENTER_VERTICAL);
            tv.setPadding(UIUtils.dip2Px(20), 0, 0, 0);
            tv.setTextColor(UIUtils.getColor(R.color.gray0));
            tv.setTextSize(14);
            tv.setText(UIUtils.getString(R.string.save_to_phone));
            mView.addView(tv);
            tv.setOnClickListener(v -> {
                if (mUrl.startsWith("file")) {
                    // 通过路径进行保存
                    File file = new File(Uri.parse(mUrl).getPath());
                    UIUtils.showToast(copyToDisk(file) ?
                            UIUtils.getString(R.string.save_success) :
                            UIUtils.getString(R.string.save_fail));
                    mPopupWindow.dismiss();       // 隐藏控件
                    mPopupWindow = null;
                } else {
                    ApiRetrofit.getInstance()
```

```
            .mApi
            .downloadPic(mUrl)
            .subscribeOn(Schedulers.newThread())
            .subscribe(responseBody -> {
        UIUtils.showToast(saveToDisk(responseBody) ?
UIUtils.getString(R.string.save_success) : UIUtils.getString(R.string.save_fail));
                        mPopupWindow.dismiss();   // 同上
                        mPopupWindow = null;
                });   }      );  }
    mPopupWindow = PopupWindowUtils.getPopupWindowAtLocation(
    mView, ViewGroup.LayoutParams.MATCH_PARENT,
    ViewGroup.LayoutParams.WRAP_CONTENT,
    getWindow().getDecorView().getRootView(), Gravity.BOTTOM, 0, 0);
    // 控件监听,头像显示大小,位置
    mPopupWindow.setOnDismissListener(() ->
    PopupWindowUtils.makeWindowLight(ShowBigImageActivity.this));
    PopupWindowUtils.makeWindowDark(ShowBigImageActivity.this);
}
```

完成上述代码,实现效果如图 7.19 和图 7.20 所示。

图 7.19　我的界面效果图

图 7.20　个人资料界面效果图

第二步:上传并更新头像。具体代码如 CORE0708 所示。

代码 CORE0708　上传头像

```java
private boolean copyToDisk(File file) {
    try {
        InputStream in = null;           // 输入流
        FileOutputStream out = null;     // 输出流
        try {
            in = new FileInputStream(file);
            out = new FileOutputStream(new File(
                AppConst.HEADER_SAVE_DIR,
                SystemClock.currentThreadTimeMillis() + "_header.jpg"));
            int c;
            while ((c = in.read()) != -1) {     // 判断输出流
                out.write(c);     // 将路径的图片发送给网络，进行上传
            }
        } catch (IOException e) {
            return false;
        } finally {
            if (in != null) {
                in.close();
            }
            if (out != null) {
                out.close();
            }
        }
        return true;
    } catch (IOException e) {
        return false;
    }
}
private boolean saveToDisk(ResponseBody body) {
    try {
        InputStream in = null;
        FileOutputStream out = null;
        try {
            in = body.byteStream();    // 将返回信息转化为输入流
            out = new FileOutputStream(new File(AppConst.HEADER_SAVE_DIR,
                SystemClock.currentThreadTimeMillis() + "_header.jpg"));
            int c;
            while ((c = in.read()) != -1) {
                out.write(c);      // 生成的数组发送给服务器
```

```
            }
        } catch (IOException e) {
            return false;
        } finally {
            if (in != null) {
                in.close();      }
            if (out != null) {
                out.close();
            }            }
            return true;
    } catch (IOException e) {
        return false;
    }}
```

第三步：修改个人昵称。具体代码如 CORE0709 所示。

代码 CORE0709　修改个人昵称

```
public class ChangeMyNameActivity extends BaseActivity {
//Activity 中用到的控件
@Bind(R.id.btnToolbarSend)
Button mBtnToolbarSend;
@Bind(R.id.etName)
EditText mEtName;
// 显示界面
@Override
public void initView() {
    mBtnToolbarSend.setText(UIUtils.getString(R.string.save));
    mBtnToolbarSend.setVisibility(View.VISIBLE);
UserInfo userInfo = DBManager.getInstance().getUserInfo(UserCache.getId());
    // 个人信息列表
    if (userInfo != null)
        mEtName.setText(userInfo.getName());
    // 得到姓名
    mEtName.setSelection(mEtName.getText().toString().trim().length());
}
@Override
public void initListener() {
    mBtnToolbarSend.setOnClickListener(v -> changeMyName());
    mEtName.addTextChangedListener(new TextWatcher() {
```

```java
            @Override
            public void beforeTextChanged(CharSequence s, int start, int count, int after) {
                /*
                在 s 字符串中，从 start 位置开始的 count 个字符即将被长度为
                after 的新文本所取代。
                */
            }
    @Override
    public void onTextChanged(CharSequence s, int start, int before, int count) {
    //s 字符串中，从 start 位置开始的 count 个字符刚刚取代了长度为 before 的旧文本
                if (mEtName.getText().toString().trim().length() > 0) {
                    mBtnToolbarSend.setEnabled(true);
                } else {
                    mBtnToolbarSend.setEnabled(false);
                }            }
    @Override
    public void afterTextChanged(Editable s) {
            //s 字符串的某个地方已经被改变
            }        });        }
    private void changeMyName() {
        // 改变昵称
        showWaitingDialog(UIUtils.getString(R.string.please_wait));
        String nickName = mEtName.getText().toString().trim();
        // 将修改的昵称传入云服务器
        ApiRetrofit.getInstance().setName(nickName)
                .subscribeOn(Schedulers.io())
                .observeOn(AndroidSchedulers.mainThread())
                .subscribe(setNameResponse -> {
                hideWaitingDialog();
        if (setNameResponse.getCode() == 200) {    // 与服务器是否连接成功
    Friend friend = DBManager.getInstance().getFriendById(UserCache.getId());
                if (friend != null) {
                    friend.setName(nickName);
                    friend.setDisplayName(nickName);
                    DBManager.getInstance().saveOrUpdateFriend(friend);
    // 从云数据库中取得信息，并将原来的替换
    BroadcastManager.getInstance(ChangeMyNameActivity.this).sendBroadcast(
```

```
        AppConst.CHANGE_INFO_FOR_ME);
BroadcastManager.getInstance(ChangeMyNameActivity.this).sendBroadcast(
    AppConst.CHANGE_INFO_FOR_CHANGE_NAME);
                    }
                        finish();
                }
        }, this::loadError);
}
private void loadError(Throwable throwable) {
    hideWaitingDialog();         // 错误提示信息
    LogUtils.sf(throwable.getLocalizedMessage());
}
@Override
protected BasePresenter createPresenter() {
    return null;
}
@Override
protected int provideContentViewId() {
    return R.layout.activity_change_name;     //Activity 的界面下信息
}}
```

第四步：查看个人二维码。具体代码如 CORE0710 所示。

代码 CORE0710　查看二维码

```
public class QRCodeCardActivity extends BaseActivity {
//Activity 中用到的控件
private UserInfo mUserInfo;
private String mGroupId;
@Bind(R.id.ivHeader)
ImageView mIvHeader;
@Bind(R.id.ngiv)
LQRNineGridImageView mNgiv;
@Bind(R.id.tvName)
TextView mTvName;
@Bind(ivCard)
ImageView mIvCard;
@Bind(R.id.tvTip)
TextView mTvTip;
```

```java
@Override
public void init() {
    mGroupId = getIntent().getStringExtra("groupId");
// 从我的界面中传来的 ID
}
@Override
public void initView() {
    mTvTip.setText(UIUtils.getString(R.string.qr_code_card_tip));
}
public void initData() {
if (TextUtils.isEmpty(mGroupId)) {
    // 获取的 ID 与云数据库中的 ID 相同,则显示对应的二维码
    mUserInfo = DBManager.getInstance().getUserInfo(UserCache.getId());
    if (mUserInfo != null) {
Glide.with(this).load(mUserInfo.getPortraitUri()).centerCrop().into(mIvHeader);
        mTvName.setText(mUserInfo.getName());
        setQRCode(AppConst.QrCodeCommon.ADD + mUserInfo.getUserId());
    }
} else {
    // 如果不对应,那么不显示二维码图片
    mNgiv.setVisibility(View.VISIBLE);
    mIvHeader.setVisibility(View.GONE);
    Observable.just(DBManager.getInstance().getGroupsById(mGroupId))
            .subscribeOn(Schedulers.io())
            .observeOn(AndroidSchedulers.mainThread())
            .subscribe(groups -> {
                if (groups == null)
                    return;
                // 显示当前用户名
                mTvName.setText(groups.getName());
            });
mNgiv.setAdapter(new LQRNineGridImageViewAdapter<GroupMember>() {
        @Override
// 压缩二维码图片
protected void onDisplayImage(Context context, ImageView imageView, GroupMember groupMember) {
Glide.with(context).load(groupMember.getPortraitUri()).
centerCrop().into(imageVie w);
```

```java
        }           });
    List<GroupMember> groupMembers = DBManager.getInstance().
getGroupMembers(mGroupId);
    // 发送到云数据库进行验证
            mNgiv.setImagesData(groupMembers);
            setQRCode(AppConst.QrCodeCommon.JOIN + mGroupId);
            mTvTip.setVisibility(View.GONE);
        }   }
private void loadQRCardError(Throwable throwable) {
    LogUtils.sf(throwable.getLocalizedMessage());   // 错误提示信息
}
@Override
protected BasePresenter createPresenter() {
    return null;
}
@Override
protected int provideContentViewId() {
    return R.layout.activity_qr_code_card;   //Activity 的布局文件
}}
```

完成上述代码,实现效果如图 7.21 所示。

图 7.21　个人二维码界面

本模块介绍了微聊项目中系统设置和用户管理功能的实现,通过本模块的学习可以了解网络连接工具的功能和使用方法,掌握 MVP 框架的用法。学习完成之后能够实现用户注册登录时与服务器之间的信息传递,并熟练运用 MVP 框架技巧。

前面学习了 MVP,首先 MVP 是从经典的 MVC 架构演变而来,不管是 MVC 还是 MVP,亦或其他架构,设计目的都是为了达到编写代码的最好效果。

技能扩展——MVC

1　MVC 简介

MVC 全名是 Model View Controller,是模型 (model) －视图 (view) －控制器 (controller) 的缩写,一种软件设计典范,用一种业务逻辑、数据、界面显示分离的方法组织代码,将业务逻辑聚集到一个部件里面,在改进和个性化定制界面及用户交互的同时,不需要重新编写业务逻辑。其中 M 层处理数据、业务逻辑等;V 层处理界面的显示结果;C 层起到桥梁的作用,来控制 V 层和 M 层通信以此来达到分离视图显示和业务逻辑层的效果。

2　MVC 执行的基本流程

首先视图接受用户输入请求,然后将请求传递给控制器,控制器再调用某个模型来处理用户的请求,在控制器的控制下,再将处理后的结果交给某个视图进行格式化输出给用户。另外,View 是可以直接访问 Model 来进行数据的处理,具体流程如图 7.22 所示。

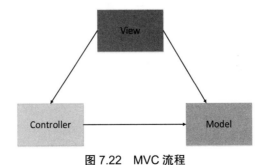

图 7.22　MVC 流程

3　MVC 的优点和缺点

如表 7.5 所示为 MVC 的优缺点。

表 7.5　MVC 优缺点

优点	缺点
耦合性低	没有明确的定义
重用性高	不适合小型，中等规模的应用程序
生命周期成本低	增加系统结构和实现的复杂性
部署快	视图与控制器之间过于紧密的连接
可维护性高	视图对模型的低效率访问
有利软件工程化管理	一般高级的界面工具或构造器不支持模式

4　MVC 的应用

MVC 在使用的过程中有固定目录结构，如图 7.23 所示是 MVC 的目录结构。

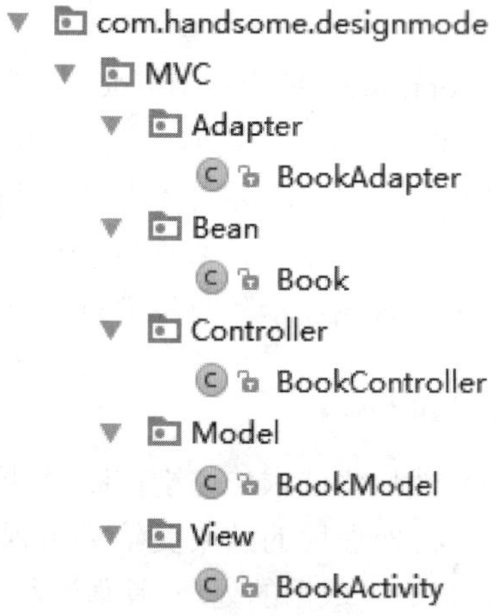

```
▼ com.handsome.designmode
    ▼ MVC
        ▼ Adapter
            © BookAdapter
        ▼ Bean
            © Book
        ▼ Controller
            © BookController
        ▼ Model
            © BookModel
        ▼ View
            © BookActivity
```

图 7.23　MVC 目录结构

根据上图可知道，项目的实体类包含书名和图片信息。具体代码如 CORE0711 所示。

代码 CORE0711　　实体类：对数据对象的封装

```java
public class Book {
    // 书名
    private String name;
    // 书的图片
    private int image;
```

```java
    public Book(String name, int image) {
        this.name = name;
        this.image = image;
    }
    public String getName() {
        return name;
    }
    public void setName(String name) {
        this.name = name;
    }
    public int getImage() {
        return image;
    }
    public void setImage(int image) {
        this.image = image;
    }
}
```

模型类（Model 层）：通常是对本地数据库的操作或者是通过网络请求获取网络数据的操作。具体代码如 CORE0712 所示。

代码 CORE0712　模型层

```java
private static List<Book> list = new ArrayList<>();
/**
 * 模拟本地数据库
 */
static {
    list.add(new Book("Java 从入门到精通 ", R.drawable.java));
    list.add(new Book("Android 从入门到精通 ", R.drawable.android));
    list.add(new Book("Java 从入门到精通 ", R.drawable.java));
    list.add(new Book("Android 从入门到精通 ", R.drawable.android));
}
/**
 * 添加书本
 * @param name
 * @param image
 */
public void addBook(String name, int image) {
```

```java
public class BookModel {
        list.add(new Book(name, image));
    }
    /**
     * 删除书本
     */
    public void deleteBook( ) {
        list.remove(list.size() - 1);
    }
    /**
     * 查询数据库所有书本
     * @return
     */
    public List<Book> query() {
        return list;
    }
}
```

控制器（Controller 层）：处理业务逻辑，调用模型层的操作，并对外暴露接口。具体代码如 CORE0713 所示。

代码 CORE0713　控制器层

```java
public class BookController {
    private BookModel mode;
    public BookController() {
        mode = new BookModel();
    }
    /**
     * 添加书本
     * @param listener
     */
    public void add(onAddBookListener listener) {
        mode.addBook("JavaWeb 从入门到精通 ", R.drawable.javaweb);
        if (listener != null) {
            listener.onComplete();
        }
    }
    /**
```

```java
 * 删除书本
 * @param listener
 */
public void delete(onDeleteBookListener listener) {
    if(mode.query().isEmpty()){
        return;
    }else{
        mode.deleteBook();
    }
    if (listener != null) {
        listener.onComplete();
    }
}
/**
 * 查询所有书本
 * @return
 */
public List<Book> query() {
    return mode.query();
}
/**
 * 添加成功的回调接口
 */
public interface onAddBookListener {
    void onComplete();
}
/**
 * 删除成功的回调接口
 */
public interface onDeleteBookListener {
    void onComplete();
}
}
```

视图（View 层）：操作 Controller 获取 List 数据填充到 ListView 中，同时可以添加书本和删除书本。具体代码如 CORE0714 所示。

代码 CORE0714　视图层

```java
public class BookActivity extends AppCompatActivity implements View.OnClickListener {
    private BookController bookController;
    private ListView lv_book;
    private List<Book> list;
    private BookAdapter adapter;
    private Button bt_add, bt_delete;
    @Override
    protected void onCreate(Bundle savedInstanceState) {
        super.onCreate(savedInstanceState);
        setContentView(R.layout.activity_book);
        lv_book = (ListView) findViewById(R.id.lv);
        bt_add = (Button) findViewById(R.id.bt_add);
        bt_delete = (Button) findViewById(R.id.bt_delete);
        bt_add.setOnClickListener(this);
        bt_delete.setOnClickListener(this);
        bookController = new BookController();
        list = bookController.query();
        adapter = new BookAdapter(this, list);
        lv_book.setAdapter(adapter);
    }
    @Override
    public void onClick(View v) {
        switch (v.getId()) {
            // 添加书本按钮
            case R.id.bt_add:
                bookController.add(new BookController.onAddBookListener() {
                    @Override
                    public void onComplete() {
                        adapter.notifyDataSetChanged();
                    }
                });
                break;
            // 删除书本按钮
            case R.id.bt_delete:
                bookController.delete(new BookController.onDeleteBookListener() {
```

```
        @Override
        public void onComplete() {
            adapter.notifyDataSetChanged();
        }
    });
break;             }   }}
```

通过以上代码实现如图 7.24 所示效果。

图 7.24　界面效果

Model	模型	Presenter	提出者
Return	返回	Login	登录
Throw	丢	Code	代码
Watcher	观察者	Desert	沙漠
Length	长度	Dismiss	解雇

一、选择题

1. 通过 LocationManager 的学习，选出获取纬度的方法（　　）。
   A. getLatitude()　　　　　　　　　　B. getLongitude()
   C. getSystemService()　　　　　　　D. getLocationURL()

2. LocationManager 所具有的功能描述不正确的是（　　）。
   A. 查询上一个已知位置
   B. 注册/注销来自某个 LocationProvider 的周期性的位置更新
   C. 注册/注销接近某个坐标时对一个已定义 Intent 的触发
   D. 帮助用户获取非常精确的位置信息

3. 位置提供器中网络定位精度稍差，耗电较少的是（　　）。
   A. GPS_PROVIDER
   B. NETWORK_PROVIDER
   C. Context.LOCATION_SERVICE
   D. ACCESS_COARSE_LOCATION

4. 用于访问 GPS 定位权限的是（　　）。
   A. <uses-permission android:name="android.permission.ACCESS_FINE_LOCATION"></uses-permission>
   B. <uses-permission android:name="android.permission.ACCESS_WIFI_STATE"></uses-permission>
   C. <uses-permission android:name="android.permission.ACCESS_NETWORK_STATE"></uses-permission>
   D. <uses-permission android:name="android.permission.CHANGE_WIFI_STATE"></uses-permission>

5. 在设备位置发生改变的时候获取到最新位置信息的方法是（　　）。
   A. requestLocationUpdates()　　　　B. getLastKnownLocation()
   C. getSystemService()　　　　　　　D. getApplication()

二、填空题

1. Retrofit 是一个可以用于_____和_____的网络库，它将自己开发的底层的代码和细节都封装了起来。
2. 使用 Retrofit 的步骤总共有_____步，其中第五步是_____。
3. MVP 的 M 指_____，V 指_____，P 指_____。
4. Presenter 处理 View 路由发送的用户操作，将其转换成相对的_____，传递给_____来做数据操作。

5. MVC 的优点是_____、_____、_____、_____、_____、_____。

三、上机题

1. 编写代码实现短信验证码注册。
2. 编写以 MVC 为框架的增加和减少。

模块二　好友及群组

通过好友以及群组模块的实现,了解 RxJava 技术的使用方法,熟悉使用函数式编程的基本方法,掌握编写添加好友的技能,具备使用所学技术开发好友群组的能力。在任务实现过程中:

● 了解 RxJava 技术的使用。
● 熟悉使用函数式编程的基本方法。
● 掌握编写添加好友实现技能。
● 具备开发好友群组的能力。

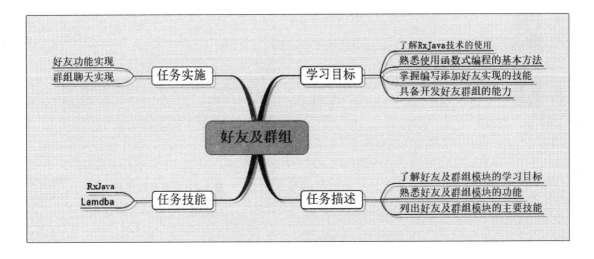

聊天软件中好友的添加和群组的创建是必不可少的,在微聊项目中通过搜索功能,输入账号或者手机号添加好友,发送验证申请,等对方通过验证后将其加入到好友列表中。该项目还具有群聊功能,好友之间可以创建群组,极大地方便了用户之间的交流。

【功能描述】

本模块将实现微聊中的添加好友和群组会话模块。

- 实现添加好友功能。
- 设置好友备注。
- 删除好友功能。
- 发起群聊或添加进群。
- 解散群聊功能或选择退出群聊。

【基本框架】

基本框架如图 8.1 至图 8.4 所示。

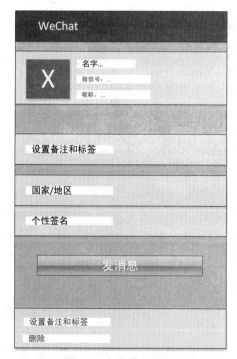

图 8.1　添加好友界面　　　　　　图 8.2　好友资料界面

项目三　微聊

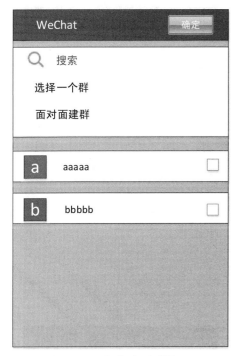

图 8.3　添加好友进群界面

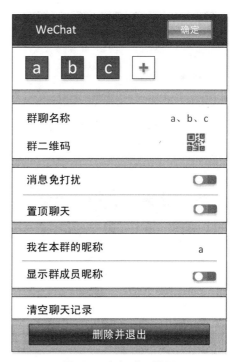

图 8.4　发起群聊界面

通过本模块的学习,将以上的框架图转换成图 8.5 至图 8.8 所示效果。

图 8.5　搜索添加好友界面

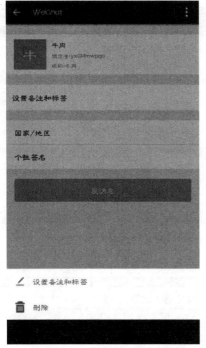

图 8.6 好友详情界面

图 8.7 添加好友进群界面

图 8.8 发起群聊界面

技能点一 RxJava

RxJava 是一个基于事件订阅异步执行的一个类库。为什么一个 Android 项目启动会联系到 RxJava 呢？因为在 RxJava 的使用得到广泛的认可，又是基于 Java 的语言。以下主要来讲解 RxJava。

1 RxJava 简介

RxJava 一个在 Java VM 上使用可观测的序列来组成异步、基于事件的程序的库。它的本质可以压缩为异步这一个词，它就是一个实现异步操作的库，已经被越来越多的人使用，Rxjava 之所以受欢迎，是因为它的优势是简洁。异步操作很关键的一点是程序的简洁性，因为在调度过程比较复杂的情况下，异步代码经常会既难写也难被读懂。RxJava 的简洁的与众不同支出在于，随着程序逻辑变得越来越复杂，它依然能够保持简洁。

2 RxJava 的观察者模式

1. 观察者模式

RxJava 的异步实现，是通过一种扩展的观察模式来实现的。观察者模式面向的需求是：A 对象（观察者）对 B 对象（被观察者）的某种变化高度敏感，需要在 B 变化的一瞬间做出反应。程序的观察者模式是采用注册（Register）或者称为订阅（Subscribe）的方式，告诉被观察者：我需要你的某某状态，你要在它变化的时候通知我。Android 开发中一个比较典型的例子是点击监听器 OnClickListener。对设置 OnClickListener 来说，View 是被观察者，OnClickListener 是观察者，二者通过 setOnClickListener() 方法达成订阅关系。订阅之后用户点击按钮的瞬间，Android Framework 就会将点击事件发送给已经注册的 OnClickListener。采取这样被动的观察方式，既省去了反复检索状态的资源消耗，也能够得到最高的反馈速度。当然，这也可以随意定制自己程序中的观察者和被观察者。

2. RxJava 的观察者模式

在 RxJava 中有四个基本概念：Observable(可观察者，即被观察者)、Observer (观察者)、subscribe (订阅)、事件。Observable 和 Observer 通过 subscribe() 方法实现订阅关系，从而 Observable 可以在需要的时候发出事件来通知 Observer。与传统观察者模式不同，RxJava 的事件回调方法除了普通事件 onNext()（相当于 onClick()/onEvent()）之外，还定义了两个特殊的事件分别是：onCompleted() 和 onError()。

● onCompleted(): 事件队列完结。RxJava 不仅把每个事件单独处理，还会把它们看做一个队列。RxJava 规定，当不会再有新的 onNext() 发出时，需要触发 onCompleted() 方法作为标志。

- onError(): 事件队列异常。在事件处理过程中出异常时，onError() 会被触发，同时队列自动终止，不允许再有事件发出。

注意：在一个正确运行的事件序列中，onCompleted() 和 onError() 有且只有一个，并且是事件序列中的最后一个。需要注意的是 onCompleted() 和 onError() 二者也是互斥的，即在队列中调用了其中一个，就不再调用另一个。

RxJava 的观察者模式如图 8.9 所示。

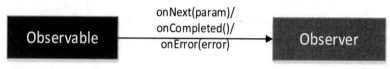

图 8.9　RxJava 的观察者模式

3. RxJava 基本实现

根据以上概念，RxJava 的基本实现主要有三点。具体如下所示。

（1）创建 Observer

Observer 即观察者，它决定事件触发时将有怎样的行为，是非常重要的。RxJava 中的 Observer 接口的实现方式如下所示。

```java
Observer<String> observer = new Observer<String>() {
    @Override
    public void onNext(String s) {
        Log.d(tag, "Item: " + s);
    }
    @Override
    public void onCompleted() {
        Log.d(tag, "Completed!");
    }
    @Override
    public void onError(Throwable e) {
        Log.d(tag, "Error!");
    }
};
```

除了 Observer 接口之外，RxJava 还内置了一个实现 Observer 的抽象类：Subscriber。Subscriber 对 Observer 接口进行了一些扩展，基本使用方式是完全一样的。具体如下所示。

```java
Subscriber<String> subscriber = new Subscriber<String>() {
    @Override
    public void onNext(String s) {
```

```
            Log.d(tag, "Item: " + s);
        }
        @Override
        public void onCompleted() {
            Log.d(tag, "Completed!");
        }
        @Override
        public void onError(Throwable e) {
            Log.d(tag, "Error!");
        }
    };
```

不仅基本使用方式一样，实质上在 RxJava 的 Subscribe 过程中，Observer 也总是会先被转换成一个 Subscriber 再使用。所以如果只想使用基本功能，选择 Observer 和 Subscriber 是完全一样的。它们的区别对于使用者来说主要有以下两点：

A.onStart(): 这是 Subscriber 增加的方法。会在 subscribe 刚开始，而事件还未发送之前被调用，可以用于做一些准备工作，例如数据的清零或重置。这是一个可选方法，默认情况下它的实现为空。需要注意的是，如果对准备工作的线程有要求（例如弹出一个显示进度的对话框，这必须在主线程执行），onStart() 就不适用了，因为总是在 Subscribe 所发生的线程被调用，而不能指定线程。要在指定的线程中做准备工作，可以使用 doOnSubscribe() 方法，具体可以在后面的文中看到。

B.unsubscribe(): 这是 Subscriber 所实现的另一个接口 Subscription 的方法，用于取消订阅。在这个方法被调用后，Subscriber 将不再接收事件。一般在这个方法调用前，可以使用 isUnsubscribed() 先判断一下状态。unsubscribe() 这个方法很重要，因为在 subscribe() 之后，Observable 会持有 Subscriber 的引用，这个引用如果不能及时被释放，将有内存泄露的风险。所以最好保持一个原则：要在不使用的时候尽快在合适的地方（例如 onPause()、onStop() 等方法中）调用 unsubscribe() 来解除引用关系，以避免内存泄露的发生。

（2）创建 Observable

Observable 即被观察者，决定什么时候触发事件以及触发怎样的事件。RxJava 使用 create() 方法来创建一个 Observable，并定义事件触发规则，具体如下所示。

```
Observable observable = Observable.create(new Observable.OnSubscribe<String>() {
        @Override
        public void call(Subscriber<? super String> subscriber) {
            subscriber.onNext("Hello");
            subscriber.onNext("Hi");
            subscriber.onNext("XiaoMa");
            subscriber.onCompleted();
        }});
```

可以看到，这里传入了一个 OnSubscribe 对象作为参数。OnSubscribe 会被存储在返回的 Observable 对象中，相当于一个计划表，当 Observable 被订阅的时候，OnSubscribe 的 call() 方法会自动被调用，事件序列就会依照设定依次触发（对于上面的代码，就是观察者 Subscriber 将会被调用三次 onNext() 和一次 onCompleted()）。这样，由被观察者调用了观察者的回调方法，就实现了由被观察者向观察者的事件传递，即观察者模式。

create() 方法是 RxJava 最基本的创造事件序列的方法。基于这个方法，RxJava 还提供了一些方法用来快捷创建事件队列，例如：

just（T…）将传入的参数依次发出来。

```
Observable observable = Observable.just("Hello", "Hi", "XiaoMa");
// 将会依次调用：
// onNext("Hello");
// onNext("Hi");
// onNext("XiaoMa");
// onCompleted();
```

from(T[]) / from(Iterable<? extends T>)：将传入的数组或 Iterable 拆分成具体对象后，依次发送出来。

```
String[] words = {"Hello", "Hi", "XiaoMa"};
Observable observable = Observable.from(words);
// 将会依次调用：
// onNext("Hello");
// onNext("Hi");
// onNext("XiaoMa");
// onCompleted();
```

这里介绍的 just（T…）和 from（T[]）和之前的 create(OnSubscribe) 是等价的。

（2）Subscribe（订阅）

创建了 Observable 和 Observer 之后，再用 subscribe() 方法将它们联结起来，整条链子就可以工作了。具体如下所示：

```
observable.subscribe(observer);
// 或者：
observable.subscribe(subscriber);
```

Observable.subscribe(Subscriber) 的内部实现的核心代码如下所示：

```
// 注意：这不是 subscribe() 的源码，而是将源码中与性能、兼容性、
// 扩展性有关的代码剔除后的核心代码。
```

```
    public Subscription subscribe(Subscriber subscriber) {
        subscriber.onStart();
        onSubscribe.call(subscriber);
        return subscriber;
    }
```

从以上代码可以看出,subscriber()做了三件事:

i. 调用 Subscriber.onStart()。这个方法在前面已经介绍过,是一个可选的准备方法。

ii. 调用 Observable 中的 OnSubscribe.call(Subscriber)。在这里,事件发送的逻辑开始运行。从这也可以看出,在 RxJava 中,Observable 并不是在创建的时候就立即开始发送事件,而是在它被订阅的时候,即当 subscribe() 方法执行的时候开始发送事件。

iii. 将传入的 Subscriber 作为 Subscription 返回。这是为了方便 unsubscribe()。

拓展:在 Android 项目的开发过程中,经常要用到一些类库来帮助项目的开发。其实用于 Android 开发的免费类库和工具集合还有很多,开发过程中如果需要,只需在网上下载即可。扫描右侧二维码了解一些免费的类库和工具集合。

通过对以上技能点的学习,下面将实现本项目好友群组模块中一些具体功能。

1. 好友

聊天软件顾名思义就是人与人之间用来交流的工具,那么添加好友成为聊天软件必不可少的功能,以下是实现好友功能的具体步骤。

第一步:通过手机号或者用户名查询好友。具体代码如 CORE0801 所示。

代码 CORE0801　　查询好友

```
public class SearchUserAtPresenter extends BasePresenter<ISearchUserAtView> {
    // 查询好友的方式
    public SearchUserAtPresenter(BaseActivity context) {
        super(context);
    }
    // 得到输入框的内容
    public void searchUser() {
        String content = getView().getEtSearchContent().getText().toString().trim();
        // 判断输入的内容
        if (TextUtils.isEmpty(content)) {
            UIUtils.showToast(UIUtils.getString(R.string.content_no_empty));
            return;
```

```java
        }
        // 显示等待窗口
        mContext.showWaitingDialog(UIUtils.getString(R.string.please_wait));
        if (RegularUtils.isMobile(content)) {
            // 通过输入的电话号码查找好友
            ApiRetrofit.getInstance().getUserInfoFromPhone(AppConst.REGION, content)
                    .subscribeOn(Schedulers.io())
                    .observeOn(AndroidSchedulers.mainThread())
                    .subscribe(getUserInfoByPhoneResponse -> {
                        mContext.hideWaitingDialog();
                        if (getUserInfoByPhoneResponse.getCode() == 200) {
                            // 通过网络接口去查询手机号对应的人
                            GetUserInfoByPhoneResponse.ResultEntity result =
                                    getUserInfoByPhoneResponse.getResult();
                            // 查询返回的信息
                            UserInfo userInfo = new UserInfo(result.getId(), result.getNickname(),
                                    Uri.parse(result.getPortraitUri()));
                            Intent intent = new Intent(mContext, UserInfoActivity.class);
                            intent.putExtra("userInfo", userInfo);
                            mContext.jumpToActivity(intent);
                        } else {
                            getView().getRlNoResultTip().setVisibility(View.VISIBLE);
                            getView().getLlSearch().setVisibility(View.GONE);
                        }
                    }, this::loadError);
        } else {
            // 通过用户名查找好友
            ApiRetrofit.getInstance().getUserInfoById(content)
                    .subscribeOn(Schedulers.io())
                    .observeOn(AndroidSchedulers.mainThread()).subscribe(
                    getUserInfoByIdResponse -> {mContext.hideWaitingDialog();
                        if (getUserInfoByIdResponse.getCode() == 200) {
                            // 通过网络接口查询用户名对应的人
                            GetUserInfoByIdResponse.ResultEntity result =
                                    getUserInfoByIdResponse.getResult();
                            // 查询返回的信息
                            UserInfo userInfo = new UserInfo(result.getId(),
```

```
                    result.getNickname(), Uri.parse(result.getPortraitUri()));
                Intent intent = new Intent(mContext, UserInfoActivity.class);
                intent.putExtra("userInfo", userInfo);
                mContext.jumpToActivity(intent);
            } else {
                getView().getRlNoResultTip().setVisibility(View.VISIBLE);
                getView().getLlSearch().setVisibility(View.GONE);
            }
        }, this::loadError);
    }  }
    private void loadError(Throwable throwable) {
        mContext.hideWaitingDialog();      // 隐藏等待提示框
        LogUtils.sf(throwable.getLocalizedMessage());
}}
```

完成上述代码，实现效果如图 8.10 所示。

图 8.10　查询并添加好友界面

第二步：根据搜索结果发起添加好友请求。具体代码如 CORE0802 所示。

代码 CORE0802　发起添加好友请求

```java
public class NewFriendAtPresenter extends BasePresenter<INewFriendAtView> {
    // 初始化列表
    private List<UserRelationshipResponse.ResultEntity> mData = new ArrayList<>();
    private LQRAdapterForRecyclerView<UserRelationshipResponse.ResultEntity>
        mAdapter;
    public NewFriendAtPresenter(BaseActivity context) {
        super(context);
    }
    // 判断是否连接网络
    public void loadNewFriendData() {
        if (!NetUtils.isNetworkAvailable(mContext)) {
            UIUtils.showToast(UIUtils.getString(R.string.please_check_net));
            return;
        }
        loadData();
        setAdapter();
    }
    private void loadData() {
        // 将内容进行网络请求
        ApiRetrofit.getInstance().getAllUserRelationship()
                .subscribeOn(Schedulers.io())
                .observeOn(AndroidSchedulers.mainThread())
                .subscribe(userRelationshipResponse -> {
                    if (userRelationshipResponse.getCode() == 200) {
                        // 请求回传的数据列表
                        List<UserRelationshipResponse.ResultEntity>
                        result = userRelationshipResponse.getResult();
                        // 分析回传数据
                        if (result != null && result.size() > 0) {
                            for (int i = 0; i < result.size(); i++) {
                                UserRelationshipResponse.ResultEntity re = result.get(i);
                                if (re.getStatus() == 10) {
                                    // 是我发起的添加好友请求
                                    result.remove(re);
                                    i--; } } }
                            if (result != null && result.size() > 0) {
```

```java
                    // 将请求信息显示在请求列表中
                    getView().getLlHasNewFriend().setVisibility(View.VISIBLE);
                        mData.clear();
                        mData.addAll(result);
                        if (mAdapter != null)
                            mAdapter.notifyDataSetChangedWrapper();
                    } else {
                        getView().getLlNoNewFriend().setVisibility(View.VISIBLE);
                    }
                } else {
Observable.error(new ServerException(UIUtils.getString(R.string.load_error)));
                }
            }, this::loadError);          }
private void setAdapter() {
    // 构建适配器
    if (mAdapter == null) {
    mAdapter = new LQRAdapterForRecyclerView<UserRelationshipResponse.
        ResultEntity>(mContext, mData, R.layout.item_new_friends) {
            @Override
public void convert(LQRViewHolderForRecyclerView helper,
        UserRelationshipResponse.ResultEntity item, int position) {
                // 通过对 item 的判断,将请求信息分类
                ImageView ivHeader = helper.getView(R.id.ivHeader);
                helper.setText(R.id.tvName, item.getUser().getNickname())
                    .setText(R.id.tvMsg, item.getMessage());
                if (item.getStatus() == 20) {// 已经是好友
                    helper.setViewVisibility(R.id.tvAdded, View.VISIBLE)
                        .setViewVisibility(R.id.tvWait, View.GONE)
                        .setViewVisibility(R.id.btnAck, View.GONE);
                } else if (item.getStatus() == 11) {// 别人发来的添加好友请求
                    helper.setViewVisibility(R.id.tvAdded, View.GONE)
                        .setViewVisibility(R.id.tvWait, View.GONE)
                        .setViewVisibility(R.id.btnAck, View.VISIBLE);
                } else if (item.getStatus() == 10) {// 我发起的添加好友请求
                    helper.setViewVisibility(R.id.tvAdded, View.GONE)
                        .setViewVisibility(R.id.tvWait, View.VISIBLE)
                        .setViewVisibility(R.id.btnAck, View.GONE);
```

```
            }
            String portraitUri = item.getUser().getPortraitUri();
            if (TextUtils.isEmpty(portraitUri)) {
                // 从云数据库中查询用户名
                portraitUri = DBManager.getInstance().getPortraitUri(
                    item.getUser().getNickname(), item.getUser().getId());
            }
            // 动态显示信息
            Glide.with(mContext).load(portraitUri).centerCrop().into(ivHeader);
            helper.getView(R.id.btnAck).setOnClickListener(v ->
                agreeFriends(item.getUser().getId(), helper));        };}
    // 填充适配器
    getView().getRvNewFriend().setAdapter(mAdapter);
}
```

第三步：根据搜索结果查看好友个人信息。具体代码如 CORE0803 所示。

代码 CORE0803　查看好友个人信息

```
public class UserInfoActivity extends BaseActivity {
//Activity 中用到的控件和一些数据的声明
UserInfo mUserInfo;
@Bind(R.id.ivHeader)
ImageView mIvHeader;
@Bind(R.id.tvName)
TextView mTvName;
@Bind(R.id.ivGender)
ImageView mIvGender;
@Bind(R.id.tvAccount)
TextView mTvAccount;
@Bind(R.id.tvNickName)
TextView mTvNickName;
@Bind(R.id.tvArea)
TextView mTvArea;
@Bind(R.id.tvSignature)
TextView mTvSignature;
private Friend mFriend;
// 来自查找好友界面传输的用户信息
@Override
```

```java
public void init() {
    Intent intent = getIntent();
    mUserInfo = intent.getExtras().getParcelable("userInfo");
    registerBR();
}
// 查询是否为好友
@Override
public void initData() {
mFriend = DBManager.getInstance().getFriendById(mUserInfo.getUserId());
Glide.with(this).load(DBManager.getInstance().getPortraitUri(mUserInfo)).
centerCrop().into(mIvHeader);
// 将所有信息填入个人信息的界面中
mTvAccount.setText(UIUtils.getString(R.string.my_chat_account,
mUserInfo.getUserId()));
mTvName.setText(mUserInfo.getName());
if (DBManager.getInstance().isMyFriend(mFriend.getUserId())) {// 我的朋友
        String nickName = mFriend.getDisplayName();
        mTvName.setText(nickName);
        if (TextUtils.isEmpty(nickName)) {
        mTvNickName.setVisibility(View.INVISIBLE);
        } else {
        mTvNickName.setText(UIUtils.getString(R.string.nickname_colon,
        mFriend.getName()));
        }      }
@Override
protected BasePresenter createPresenter() {
    return null;      }
@Override
protected int provideContentViewId() {
    return R.layout.activity_user_info;        // 好友信息界面
}}
```

完成上述代码,实现如图 8.11 所示效果。

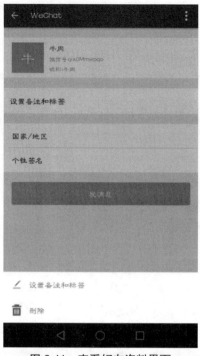

图 8.11 查看好友资料界面

第四步:在添加好友完成后,为了后期方便查找设置好友备注。具体代码如 CORE0804 所示。

代码 CORE0804 设置备注

```
public class SetAliasActivity extends BaseActivity {
// 初始化变量
private String mFriendId;
private Friend mFriend;
//Activity 中用到的控件
@Bind(R.id.btnToolbarSend)
Button mBtnToolbarSend;
@Bind(R.id.etAlias)
EditText mEtAlias;
// 查询好友界面传送的用户名
@Override
public void init() {
    mFriendId = getIntent().getStringExtra("userId");
}
// 查询是否为好友
@Override
```

```java
public void initView() {
    if (TextUtils.isEmpty(mFriendId)) {
        finish();
        return;
    }
    mBtnToolbarSend.setVisibility(View.VISIBLE);
    mBtnToolbarSend.setText(UIUtils.getString(R.string.complete));
}
@Override
public void initListener() {
    mEtAlias.addTextChangedListener(new TextWatcher() {
        @Override
        public void beforeTextChanged(CharSequence s, int start, int count, int after) {
            // 在 s 字符串中,从 start 位置开始的 count 个字符即将被长度为 after 的新文本
            // 所取代
        }
        @Override
        public void onTextChanged(CharSequence s, int start, int before, int count) {
            mBtnToolbarSend.setEnabled(mEtAlias.getText().toString().trim().length() > 0 ? true : false);
            //s 字符串中,从 start 位置开始的 count 个字符刚刚取代了长度为
            //before 的旧文本
        }
        @Override
        public void afterTextChanged(Editable s) {
            //s 字符串的某个地方已经被改变
        }
    });
    // 点击完成按钮的点击事件
    mBtnToolbarSend.setOnClickListener(v -> {
        String displayName = mEtAlias.getText().toString().trim();
        if (TextUtils.isEmpty(displayName)) {
            UIUtils.showToast(UIUtils.getString(R.string.alias_no_empty));
            return;
        }
        // 请稍等
        showWaitingDialog(UIUtils.getString(R.string.please_wait));
        // 修改备注
```

```java
                ApiRetrofit.getInstance().setFriendDisplayName(mFriendId, displayName)
                        .subscribeOn(Schedulers.io())
                        .observeOn(AndroidSchedulers.mainThread())
                        .subscribe(setFriendDisplayNameResponse -> {
                            if (setFriendDisplayNameResponse.getCode() == 200) {
                                // 修改成功
                                UIUtils.showToast(UIUtils.getString(R.string.change_success));
                                // 更新本地好友数据库
                                mFriend.setDisplayName(displayName);
                                mFriend.setDisplayNameSpelling(PinyinUtils.getPinyin(displayName));
                                DBManager.getInstance().saveOrUpdateFriend(mFriend);
                                BroadcastManager.getInstance(SetAliasActivity.this).
                                        sendBroadcast(AppConst.UPDATE_FRIEND);
                                BroadcastManager.getInstance(SetAliasActivity.this).sendBroadcast(
                                        AppConst.CHANGE_INFO_FOR_USER_INFO);
                                finish();
                            } else {
                                // 修改失败
                                UIUtils.showToast(UIUtils.getString(R.string.change_fail));
                            }
                            hideWaitingDialog();
                        }, this::changeError);
    });      }
    private void changeError(Throwable throwable) {        // 出错打印并显示
        hideWaitingDialog();
        LogUtils.sf(throwable.getLocalizedMessage());
        UIUtils.showToast(throwable.getLocalizedMessage());
    }
    @Override
    protected int provideContentViewId() {                 //Activity 的布局文件
        return R.layout.activity_set_alias;
    }}
```

第五步：在整理好友列表时可进行好友的删除。具体代码如 CORE0805 所示。

代码 CORE0805 删除好友

```java
// 删除按钮监听
mOivDelete.setOnClickListener(v -> {
```

```java
// 点击删除,弹出提示框
showMaterialDialog(UIUtils.getString(R.string.delete_contact),
UIUtils.getString(R.string.delete_contact_content, mUserInfo.getName()),
        UIUtils.getString(R.string.delete),
        UIUtils.getString(R.string.cancel),
// 点击确定,将联系人的所有信息及聊天记录全部删除
v1 -> ApiRetrofit.getInstance()
        .deleteFriend(mUserInfo.getUserId())
        .subscribeOn(Schedulers.io())
        .observeOn(AndroidSchedulers.mainThread())
        .subscribe(deleteFriendResponse -> {
        hideMaterialDialog();
        // 点击删除后的响应
        if (deleteFriendResponse.getCode() == 200) {
        // 响应等于 200 将执行删除操作
        RongIMClient.getInstance().getConversation(
        Conversation.ConversationType.PRIVATE,
        mUserInfo.getUserId(), new RongIMClient.ResultC
            allback<Conversation>() {
      @Override
      public void onSuccess(Conversation conversation) {
          // 清除所有信息
          RongIMClient.getInstance().clearMessages(
          Conversation.ConversationType.PRIVATE,
          mUserInfo.getUserId(), new
          RongIMClient.ResultCallback<Boolean>() {
          @Override
      public void onSuccess(Boolean aBoolean) {
          // 移除所有信息
          RongIMClient.getInstance().removeConversation(
          Conversation.ConversationType.PRIVATE,
          mUserInfo.getUserId(), null);
                                }
           @Override
      public void onError(RongIMClient.ErrorCode errorCode) { } });}
             @Override
      public void onError(RongIMClient.ErrorCode errorCode) {
```

```
                    }
                });
        // 通知对方被删除 ( 把我的 id 发给对方 )
        DeleteContactMessage deleteContactMessage =
        DeleteContactMessage.obtain(UserCache.getId());
        RongIMClient.getInstance().sendMessage(
        Message.obtain(mUserInfo.getUserId(),
        Conversation.ConversationType.PRIVATE,
        deleteContactMessage), "", "", null, null);
DBManager.getInstance().deleteFriendById(mUserInfo.getUserId());
// 删除成功
UIUtils.showToast(UIUtils.getString(R.string.delete_success));
BroadcastManager.getInstance(UserInfoActivity.this).sendBroadcast(
AppConst.UPDATE_FRIEND);
                    finish();
                } else {
// 删除失败
UIUtils.showToast(UIUtils.getString(R.string.delete_fail));
                }}, this::loadError,v2 -> hideMaterialDialog());
            });}
```

第六步：加好友的方式不仅仅是搜索用户名/手机号，其中扫码加朋友是更加便捷的方法。具体代码如 CORE0806 所示。

代码 CORE0806　扫码加朋友

```
// 扫码加好友
mZxingview.setDelegate(this);
@Override
    protected void onStart() {
        super.onStart();
        // 启动扫一扫
        mZxingview.startCamera();
        mZxingview.startSpotAndShowRect();
    }
    @Override
    protected void onStop() {
        super.onStop();
        // 停止扫码
```

```java
            mZxingview.stopCamera();
        }
        @Override
        protected void onDestroy() {
            super.onDestroy();
            mZxingview.onDestroy();
        }
        @Override
        public void onActivityResult(int requestCode, int resultCode, Intent data) {
            super.onActivityResult(requestCode, resultCode, data);
            if (resultCode == ImagePicker.RESULT_CODE_ITEMS) {// 返回多张照片
                if (data != null) {
                    final ArrayList<ImageItem> images = (ArrayList<ImageItem>) data.getSerializableExtra(ImagePicker.EXTRA_RESULT_ITEMS);
                    if (images != null && images.size() > 0) {
                        // 取第一张照片
                        ThreadPoolFactory.getNormalPool().execute(new Runnable() {
                            @Override
                            public void run() {
                                // 扫描结果
                                String result = QRCodeDecoder.syncDecodeQRCode(images.get(0).path);
                                if (TextUtils.isEmpty(result)) {
                                    UIUtils.showToast(UIUtils.getString(R.string.scan_fail));
                                } else {
                                    handleResult(result);
                                }}});}}}
// 定义线程处理扫描结果
private void handleResult(String result) {
    LogUtils.sf(" 扫描结果 :" + result);
    vibrate();
    mZxingview.startSpot();
    // 添加好友
    if (result.startsWith(AppConst.QrCodeCommon.ADD)) {
        String account = result.substring(AppConst.QrCodeCommon.ADD.length());
        // 更新数据
        if (DBManager.getInstance().isMyFriend(account)) {
            // 如果存在则不需要添加
```

```java
            UIUtils.showToast(UIUtils.getString(R.string.this_account_was_your_friend));
                return;
            }
        // 不存在,需要添加
        Intent intent = new Intent(ScanActivity.this, PostScriptActivity.class);
                intent.putExtra("userId", account);
                startActivity(intent);
                finish();
            }
    // 添加朋友
    public void addFriend(String userId) {
            String msg = getView().getEtMsg().getText().toString().trim();
            ApiRetrofit.getInstance().sendFriendInvitation(userId, msg)
            .subscribeOn(Schedulers.io())
            .observeOn(AndroidSchedulers.mainThread())
            .subscribe(friendInvitationResponse -> {
            if (friendInvitationResponse.getCode() == 200) {
            // 请求发送
            UIUtils.showToast(UIUtils.getString(R.string.rquest_sent_success));
                mContext.finish();
            } else {
                UIUtils.showToast(UIUtils.getString(R.string.rquest_sent_fail));
            }
            }, this::loadError);
}
```

第七步:添加好友之后可查看新加朋友。具体代码如 CORE0807 所示。

代码 CORE0807　　查看新加朋友
| ```java
    // 同意请求
private void agreeFriends(String friendId, LQRViewHolderForRecyclerView helper){
            if (!NetUtils.isNetworkAvailable(mContext)) {
                UIUtils.showToast(UIUtils.getString(R.string.please_check_net));
                return;
            }
            // 将同意信息返回给服务器
            ApiRetrofit.getInstance().agreeFriends(friendId)
                .subscribeOn(Schedulers.io())
``` |
```

```java
            .observeOn(AndroidSchedulers.mainThread())
            .flatMap(new Func1<AgreeFriendsResponse,
            Observable<GetUserInfoByIdResponse>>(){
@Override
// 判断得到请求数据
public Observable<GetUserInfoByIdResponse> call(AgreeFriendsResponse agreeFriendsResponse) {
    if (agreeFriendsResponse != null && agreeFriendsResponse.getCode() == 200) {
        helper.setViewVisibility(R.id.tvAdded, View.VISIBLE).
        setViewVisibility(R.id.btnAck, View.GONE);
        // 添加成功
        return ApiRetrofit.getInstance().getUserInfoById(friendId);
    }
    // 添加失败
    return Observable.error(new ServerException(UIUtils.getString(
    R.string.agree_friend_fail)));
        } })
            .subscribe( -> {
// 判断好友信息的回传响应
    if (getUserInfoByIdResponse != null &&
    getUserInfoByIdResponse.getCode() ==200) {
    GetUserInfoByIdResponse.ResultEntity result =
    getUserInfoByIdResponse.getResult();
    UserInfo userInfo = new UserInfo(UserCache.getId(),
    result.getNickname(),Uri.parse(result.getPortraitUri()));
    // 更新头像信息
    if (TextUtils.isEmpty(userInfo.getPortraitUri().toString())) {
        userInfo.setPortraitUri(Uri.parse(
        DBManager.getInstance().getPortraitUri(userInfo)));
            }
        Friend friend = new Friend(userInfo.getUserId(),
        userInfo.getName(), userInfo.getPortraitUri().toString());
        DBManager.getInstance().saveOrUpdateFriend(friend);
        UIUtils.postTaskDelay(() -> {
            // 更新本地数据
BroadcastManager.getInstance(UIUtils.getContext()).sendBroadcast(
        AppConst.UPDATE_FRIEND);
```

```
        BroadcastManager.getInstance(UIUtils.getContext()).sendBroadcast(
        AppConst.UPDATE_CONVERSATIONS);
                }, 1000);              }
            }, this::loadError);       }
```

2. 群组

通过上边的任务，已经完成了好友的添加，如果觉得单聊不够过瘾，那么多个好友可进行建群聊天，以下是实现项目群组功能的具体步骤。

第一步：首先是将多个好友邀请进群。具体代码如 CORE0808 所示。

代码 CORE0808　拉人进群

```java
public class CreateGroupActivity extends BaseActivity<ICreateGroupAtView, CreateGroupAtPresenter> implements ICreateGroupAtView {
    // 初始化列表 list
    public ArrayList<String> mSelectedTeamMemberAccounts;
    //Activity 中用到的控件
    @Bind(R.id.btnToolbarSend)
    Button mBtnToolbarSend;
    @Bind(R.id.rvSelectedContacts)
    LQRRecyclerView mRvSelectedContacts;
    @Bind(R.id.etKey)
    EditText mEtKey;
    private View mHeaderView;
    @Bind(R.id.rvContacts)
    LQRRecyclerView mRvContacts;
    @Bind(R.id.qib)
    QuickIndexBar mQib;
    @Bind(R.id.tvLetter)
    TextView mTvLetter;
    // 接收传入的数字列
    @Override
    public void init() {
        mSelectedTeamMemberAccounts = getIntent().getStringArrayListExtra("selectedMember");
    }
    // 找到对应 ID 的控件
    @Override
    public void initView() {
```

```java
            mBtnToolbarSend.setVisibility(View.VISIBLE);
            mBtnToolbarSend.setText(UIUtils.getString(R.string.sure));
            mBtnToolbarSend.setEnabled(false);
            mHeaderView = View.inflate(this, R.layout.header_group_cheat, null);
    }
    @Override
    public void initListener() {
        // 创建群聊,并邀请成员
        mBtnToolbarSend.setOnClickListener(v -> {
            if (mSelectedTeamMemberAccounts == null) {
                mPresenter.createGroup();
            } else {
                // 添加群成员
                mPresenter.addGroupMembers();
            }
        });
    private void addMember(boolean isAddMember) {
        Intent intent = new Intent(mContext, CreateGroupActivity.class);
        // 如果是群组的话就把当前已经进群的群成员发过去
        if (isAddMember) {
            ArrayList<String> selectedTeamMemberAccounts = new ArrayList<>();
            for (int i = 0; i < mData.size(); i++) {
                selectedTeamMemberAccounts.add(mData.get(i).getUserId());
            }
            intent.putExtra("selectedMember", selectedTeamMemberAccounts);
        }
mContext.startActivityForResult(intent,SessionInfoActivity.REQ_ADD_MEMBERS);
}
        public void addGroupMember(ArrayList<String> selectedIds) {
            LogUtils.sf("addGroupMember : " + selectedIds);
            mContext.showWaitingDialog(UIUtils.getString(R.string.please_wait));
            ApiRetrofit.getInstance().addGroupMember(mSessionId, selectedIds)
                .subscribeOn(Schedulers.io())
                .observeOn(AndroidSchedulers.mainThread())
                .subscribe(addGroupMemberResponse -> {
                    if (addGroupMemberResponse != null &&
                        addGroupMemberResponse.getCode() == 200) {
```

```java
        LogUtils.sf(" 网络请求成功,开始添加群成员: ");
        // 更新群成员
        Groups groups = DBManager.getInstance().getGroupsById(mSessionId);
            for (String groupMemberId : selectedIds) {
// 更新群成员信息
UserInfo userInfo = DBManager.getInstance().getUserInfo(groupMemberId);
            if (userInfo != null) {
                GroupMember newMember = new GroupMember(mSessionId,
                                userInfo.getUserId(),
                                userInfo.getName(),
                                userInfo.getPortraitUri().toString(),
                                userInfo.getName(),
                                PinyinUtils.getPinyin(userInfo.getName()),
                                PinyinUtils.getPinyin(userInfo.getName()),
                                groups.getName(),
                                PinyinUtils.getPinyin(groups.getName()),
                                groups.getPortraitUri());
                // 保存到本地数据
                DBManager.getInstance().saveOrUpdateGroupMember(newMember);
                        LogUtils.sf(" 添加群成员成功 ");
                    }
                }
            LogUtils.sf(" 添加群成员结束 ");
            mContext.hideWaitingDialog();
            loadData();
            LogUtils.sf(" 重新加载数据 ");
            UIUtils.showToast(UIUtils.getString(R.string.add_member_success));
            }
    }, this::addMembersError);}
```

完成上述代码,实现如图 8.12 和 8.13 所示效果。

图 8.12　邀请好友进群组

图 8.13　群聊界面

第二步：邀请人进群同样也可踢人出群。具体代码如 CORE0809 所示。

代码 CORE0809　踢人出群

```java
public class RemoveGroupMemberActivity extends BaseActivity {
// 初始化数据
private String mGroupId;
private List<GroupMember> mData = new ArrayList<>();
private List<GroupMember> mSelectedData = new ArrayList<>();
//Activity 中用到的控件
@Bind(R.id.btnToolbarSend)
Button mBtnToolbarSend;
@Bind(R.id.rvMember)
LQRRecyclerView mRvMember;
// 定义适配器
private LQRAdapterForRecyclerView<GroupMember> mAdapter;
@Override
public void init() {
    mGroupId = getIntent().getStringExtra("sessionId");  // 接收传入的组号
}
@Override
```

```java
public void initView() {
    // 判断接收的组号
    if (TextUtils.isEmpty(mGroupId)) {
        finish();
        return;
    }
    // 点击删除
    mBtnToolbarSend.setText(UIUtils.getString(R.string.delete));
    mBtnToolbarSend.setVisibility(View.VISIBLE);
    mBtnToolbarSend.setBackgroundResource(R.drawable.shape_btn_delete);
    mBtnToolbarSend.setEnabled(false);
}
@Override
public void initData() {
    // 网络请求得到的列号
    List<GroupMember> groupMembers =
    DBManager.getInstance().getGroupMembers(mGroupId);
    if (groupMembers != null && groupMembers.size() > 0) {
        for (int i = 0; i < groupMembers.size(); i++) {
            GroupMember groupMember = groupMembers.get(i);
            if (groupMember.getUserId().equals(UserCache.getId())) {
                // 找到这个列对应的 ID 号,并把这个 ID 号移除
                groupMembers.remove(i);
                break;
            }
        }
        mData.clear();
        mData.addAll(groupMembers);
    }
    setAdapter();
}
// 按钮监听
@Override
public void initListener() {
    mBtnToolbarSend.setOnClickListener(v -> {
        ArrayList<String> selectedIds = new ArrayList<>(mSelectedData.size());
        // 循环获取到 ID
        for (int i = 0; i < mSelectedData.size(); i++) {
            GroupMember groupMember = mSelectedData.get(i);
```

```
                selectedIds.add(groupMember.getUserId());
        }
            // 携带将要删除的 Id 号给群组界面
            Intent data = new Intent();
            data.putStringArrayListExtra("selectedIds", selectedIds);
            setResult(Activity.RESULT_OK, data);
            finish();
    });   }
// 删除群成员
public void deleteGroupMembers(ArrayList<String> selectedIds) {
mContext.showWaitingDialog(UIUtils.getString(R.string.please_wait));
ApiRetrofit.getInstance().deleGroupMember(mSessionId, selectedIds)
        .subscribeOn(Schedulers.io())
        .observeOn(AndroidSchedulers.mainThread())
        .subscribe(deleteGroupMemberResponse -> {
    // 网络请求
    if (deleteGroupMemberResponse != null &&
        deleteGroupMemberResponse.getCode() == 200) {
                LogUtils.sf(" 网络请求成功,开始删除:");
                for (int i = 0; i < mData.size(); i++) {
                    GroupMember member = mData.get(i);
                    // 删除用户
                    if (selectedIds.contains(member.getUserId())) {
                        LogUtils.sf(" 删除用户:" + member.getUserId());
                        member.delete();
                        mData.remove(i);
                        i--;  }}
        LogUtils.sf(" 删除结束 ");
        mContext.hideWaitingDialog();
        setAdapter();
        UIUtils.showToast(UIUtils.getString(R.string.del_member_success));
        } else {
            LogUtils.sf(" 网络请求失败 ");
            mContext.hideWaitingDialog();
            UIUtils.showToast(UIUtils.getString(R.string.del_member_fail));
        }
    }, this::delMembersError);}
```

第三步：修改群昵称。具体代码如 CORE0810 所示。

代码 CORE0810　修改群昵称

```java
public class SetGroupNameActivity extends BaseActivity {
// 初始化数据
private String mGroupId;
//Activity 中用到的控件
@Bind(R.id.btnToolbarSend)
Button mBtnToolbarSend;
@Bind(R.id.etName)
EditText mEtName;
@Override
public void init() {
    mGroupId = getIntent().getStringExtra("groupId");    // 接收传入的组号
}
@Override
public void initData() {
// 通过接收的组号去请求服务器数据库，找到对应群 ID 的群组
    Observable.just(DBManager.getInstance().getGroupsById(mGroupId))
            .subscribeOn(Schedulers.io())
            .observeOn(AndroidSchedulers.mainThread())
            .subscribe(groups -> {
                if (groups != null) {
                mEtName.setText(groups.getName());
                mEtName.setSelection(groups.getName().length());
                mBtnToolbarSend.setEnabled(groups.getName().length() > 0);
                } }); }
@Override
public void initListener() {
    mEtName.addTextChangedListener(new TextWatcher() {
        @Override
// 在 s 字符串中，从 start 位置开始的 count 个字符即将被长度为
//after 的新文本所取代
public void beforeTextChanged(CharSequence s, int start, int count, int after) {
        }
        @Override
// 在 s 字符串中，从 start 位置开始的 count 个字符刚刚取代了长度为
//before 的旧文本
```

```java
        public void onTextChanged(CharSequence s, int start, int before, int count) {
            mBtnToolbarSend.setEnabled(mEtName.getText().toString().trim().length() > 0);
        }
        @Override
//s 字符串的某个地方已经被改变
        public void afterTextChanged(Editable s) {
        }
    });
    mBtnToolbarSend.setOnClickListener(v -> {
        String groupName = mEtName.getText().toString().trim();
        if (!TextUtils.isEmpty(groupName)) {
            showWaitingDialog(UIUtils.getString(R.string.please_wait));
            ApiRetrofit.getInstance().setGroupName(mGroupId, groupName)
                    .subscribeOn(Schedulers.io())
                    .observeOn(AndroidSchedulers.mainThread())
                    .subscribe(setGroupNameResponse -> {
                        if (setGroupNameResponse != null &&
                            setGroupNameResponse.getCode() == 200) {
                            Groups groups = DBManager.getInstance().getGroupsById(mGroupId);
                            if (groups != null) {
                                groups.setName(groupName);
                                groups.saveOrUpdate("groupid=?", groups.getGroupId());
                            }
                            // 发送广播
BroadcastManager.getInstance(SetGroupNameActivity.this).sendBroadcast(
AppConst.UPDATE_CONVERSATIONS);
BroadcastManager.getInstance(SetGroupNameActivity.this).sendBroadcast(
AppConst.UPDATE_CURRENT_SESSION_NAME);
                            UIUtils.showToast(UIUtils.getString(R.string.set_success));
// 设置成功,界面跳转
Intent data = new Intent();
data.putExtra("group_name", groupName);
                            setResult(RESULT_OK, data);
                            hideWaitingDialog();
                                finish();
                        } else {
                            // 等待提示框
                            hideWaitingDialog();
```

```
                    // 设置失败
                    UIUtils.showToast(UIUtils.getString(R.string.set_fail));
                }
            }, this::loadError);
        }   });   }
```

第四步：如果不是好友可通过扫码加入群组。具体代码如 CORE0811 所示。

代码 CORE0811　扫码加入群组

```
    // 扫码加好友
    mZxingview.setDelegate(this);
    @Override
protected void onStart() {
    super.onStart();
                // 启动扫一扫
                mZxingview.startCamera();
                mZxingview.startSpotAndShowRect();
            }
            @Override
            protected void onStop() {
                super.onStop();
                // 停止扫码
                mZxingview.stopCamera();
            }
            @Override
            protected void onDestroy() {
                super.onDestroy();
                mZxingview.onDestroy();
            }
            @Override
    public void onActivityResult(int requestCode, int resultCode, Intent data) {
      super.onActivityResult(requestCode, resultCode, data);
        if (resultCode == ImagePicker.RESULT_CODE_ITEMS) {// 返回多张照片
          if (data != null) {
    final ArrayList<ImageItem> images = (ArrayList<ImageItem>) data.getSerializa
                bleExtra(ImagePicker.EXTRA_RESULT_ITEMS);
                if (images != null && images.size() > 0) {
                    // 取第一张照片
```

```java
            ThreadPoolFactory.getNormalPool().execute(new Runnable() {
                @Override
                public void run() {
                    // 扫描结果
                    String result = QRCodeDecoder.syncDecodeQRCode(images.get(0).path);
                    if (TextUtils.isEmpty(result)) {
                        UIUtils.showToast(UIUtils.getString(R.string.scan_fail));
                    } else {
                        handleResult(result);
                    }}});  }  }  }}
// 定义线程处理扫描结果
private void handleResult(String result) {
            LogUtils.sf(" 扫描结果 :" + result);
            vibrate();
            mZxingview.startSpot();
            // 进群
            if (result.startsWith(AppConst.QrCodeCommon.JOIN)) {
    String groupId = result.substring(AppConst.QrCodeCommon.JOIN.length());
    // 请求数据
    if (DBManager.getInstance().isInThisGroup(groupId)) {
    // 如果在群组中，则不需要进入
    UIUtils.showToast(UIUtils.getString(R.string.you_already_in_this_group));
            return;
    } else {
    // 请求加入群组
    ApiRetrofit.getInstance().JoinGroup(groupId)
            .subscribeOn(Schedulers.io())
            .observeOn(AndroidSchedulers.mainThread())
            .filter(joinGroupResponse -> joinGroupResponse != null &&
            joinGroupResponse.getCode() == 200)
            .flatMap(new Func1<JoinGroupResponse,
            Observable<GetGroupInfoResponse>>() {
    @Override
    public Observable<GetGroupInfoResponse> call(JoinGroupResponse join-
GroupResponse) {
                return ApiRetrofit.getInstance().getGroupInfo(groupId);
            } })
```

```java
            .subscribe(getGroupInfoResponse -> {
                // 回传响应
                if (getGroupInfoResponse != null &&
                getGroupInfoResponse.getCode() == 200) {
                GetGroupInfoResponse.ResultEntity resultEntity =
                getGroupInfoResponse.get
                Result();
                // 更新本地数据
    DBManager.getInstance().saveOrUpdateGroup(new Groups(resultEntity.getId(),
                    resultEntity.getName(), null, String.valueOf(0)));
                // 将得到的信息传入查询界面
    Intent intent = new Intent(ScanActivity.this, SessionActivity.class);
        intent.putExtra("sessionId", resultEntity.getId());
        intent.putExtra("sessionType", SessionActivity.SESSION_TYPE_GROUP);
                    jumpToActivity(intent);
                        finish();
                    } else {
// 已加入群聊
    Observable.error(new ServerException(UIUtils.getString(
R.string.select_group_info_fail_please_restart_app)));
                    }
                }, this::loadError);
        }}
```

第五步：对于不满意的群可进行解散群或者退出群。具体代码如 CORE0812 所示。

代码 CORE0812   解散群，退出群

```java
public class Groups extends DataSupport implements Serializable {
    private String role;        //0 是管理员，1 是群成员
public String getRole() {
    return role;
}}
// 判断是管理员还是群成员
if (mGroups.getRole().equalsIgnoreCase("0")) {
    // 是否确定要解散该群
    tip = UIUtils.getString(R.string.are_you_sure_to_dismiss_this_group);
    quitGroupResponseObservable = ApiRetrofit.getInstance().dissmissGroup(mSessionId);
```

```java
        } else {
            // 是否退出该群,退出后将不再接受此群的消息
            tip = UIUtils.getString(R.string.you_will_never_receive_any_msg_after_quit);
            quitGroupResponseObservable = ApiRetrofit.getInstance().quitGroup(mSessionId);
        }
        // 确定或者取消,请求响应
mContext.showMaterialDialog(null, tip, UIUtils.getString(R.string.sure),
    UIUtils.getString(R.string.cancel) , v -> quitGroupResponseObservable
                .subscribeOn(Schedulers.io())
                .observeOn(AndroidSchedulers.mainThread())
                .subscribe(quitGroupResponse -> {
                    mContext.hideMaterialDialog();
                    // 返回请求响应
                    if (quitGroupResponse != null &&
                quitGroupResponse.getCode() == 200) {
                    RongIMClient.getInstance().getConversation(mConversationType,
                    mSessionId,new RongIMClient.ResultCallback<Conversation>() {
                        @Override
                        // 得到请求响应
            public void onSuccess(Conversation conversation) {
            // 清除所有消息
            RongIMClient.getInstance().clearMessages(Conversation.ConversationType.
GROUP, mSessionId, new RongIMClient.ResultCallback<Boolean>() {
            @Override
public void onSuccess(Boolean aBoolean) {
    // 退出后不接收信息
    RongIMClient.getInstance().removeConversation(mConversationType,
mSessionId, null); }
                @Override
                public void onError(RongIMClient.ErrorCode errorCode) {
                                }); }
                @Override
                public void onError(RongIMClient.ErrorCode errorCode) {
                        } });
// 更新本地数据
DBManager.getInstance().deleteGroupMembersByGroupId(mSessionId);
DBManager.getInstance().deleteGroupsById(mSessionId);
```

```
//动态更新
BroadcastManager.getInstance(mContext).sendBroadcast(
AppConst.UPDATE0_CONVERSATIONS);
BroadcastManager.getInstance(mContext).sendBroadcast(
AppConst.UPDATE_GROUP);
BroadcastManager.getInstance(mContext).sendBroadcast(
AppConst.CLOSE_CURRENT_SESSION);
    mContext.finish();
                    } else {
//显示退出失败
UIUtils.showToast(UIUtils.getString(R.string.exit_group_fail));
                    }
                }, this::quitError)
//等待弹窗
, v -> mContext.hideMaterialDialog()); }
```

本模块介绍了好友群组的实现,通过本模块的学习可以了解 RxJava 技术的使用方法,掌握使用函数式编程的基本用法,学习之后能够实现添加好友、创建群组、解散群、退群等功能。

Java 8 中引入了一些非常有特色的功能。如 Lambda 表达式、streamAPI、接口默认实现等。Lambda 表达式本质上是一种匿名方法,它既没有方法名,也没有访问修饰符和返回值类型,使用 Lambda 表达式来编写代码将会更加简洁,更加易读。

技能扩展——Lambda

1 简介

Lambda 是函数式编程中的基础部分,已经在其他编程语言(例如:Scala)中使用,但在 Java 领域中发展较慢,直到 Java 8,才开始支持 Lambda。Lambda 可以使表达更加简洁,Java 8 之前要想向方法传入一个函数逻辑,不得不去实现一个接口,Java 8 带来的 Lambda 表达式可以将函数作为参数直接传入方法内。

2　Lambda 在 Android Studio 下的环境搭建

下面首先搭建环境,因为 Android 不支持 Java 8,所以需要用到一个开源库:retolambda（下载地址:https://github.com/evant/gradle-retrolambda）,接着是下载 java8（下载地址:http://www.oracle.com/technetwork/java/javase/downloads/jdk8-downloads-2133151.html）。

1. 修改配置工程文件

首先在 app/budild.gradle 中添加如下配置:

```
jackOptions{
enabled true
}
// 设置 JDK1.8
compileOptions{
    sourceCompatibility JavaVersion.VERSION_1_8
    targetCompatibility JavaVersion.VERSION_1_8
}
// 添加插件
apply plugin: 'me.tatarka.retrolambda'
```

3　Lambda 表达式的用法

通常 Lambda 表达式有三种写法,第一种方式为"无参数+语句（代码块）":一般适用于内部类中方法无参数的情况。第二种方式为"有参数+语句":适用于匿名内部类中方法只有一个参数的情况。第三种方式为"有参数+代码块":适用于匿名内部类种方法不止一个参数的情况。详细代码如下:

1. 无参数+语句（代码块）

```
// () -> 语句 或 () -> { 代码块 }
private void threadTest()
/** 普通写法 **/
new Thread(new Runnable() {
@Override public void run() {
    }
}).start();
/** 使用 lambda 表达式写法 **/
new Thread(() -> Log.d(TAG, "this is a lambda Thread")).start();
}
```

2. 有参数 + 语句

```
// 方法参数 -> 语句 或 * 方法参数 -> { 代码块 }
 private void setOnClick() {
  /** 普通写法 **/
  findViewById(R.id.button).setOnClickListener(new View.OnClickListener() {
   @Override
   public void onClick(View v) {
    Log.d(TAG, "this is a general onClick");
   }
  });
  /** 使用 lambda 表达式写法 **/
  findViewById(R.id.button).setOnClickListener(v -> Log.d(TAG,
"this is a lambda onClick"));
 }
```

3. 有参数 + 代码块

```
//( 参数1, 参数2) -> 语句 或 ( 参数1, 参数2) -> { 代码块 }
 private void setOnChecked()
 {
CheckBox checkBox = (CheckBox) findViewById(R.id.ch//eckBox);
 /** 普通写法 **/
checkBox.setOnCheckedChangeListener(new
CompoundButton.OnCheckedChangeListener() {
 @Override
 public void onCheckedChanged(CompoundButton buttonView, boolean isChecked){
  Log.d(TAG, "this is a general onCheckedChanged");
 }
 });
 /** 使用 lambda 表达式写法 **/
 checkBox.setOnCheckedChangeListener((buttonView, isChecked) -> {
  Log.d(TAG, "this is a lambda onCheckedChanged");
  Log.d(TAG, "this is a lambda onCheckedChanged_isChecked=" + isChecked);
 });
 }
```

Group	群组	Compress	压缩
List	序列	Asynchronous	异步
Operation	操作	Observe	观察
Rely	依赖	Example	实例
Abstract	抽象	Subscription	订阅

一、选择题

1. 下列不属于使用 Retrofit 的步骤是（　　）。

A. 添加 Retrofit 库的依赖

B. 创建用于描述网络请求的接口

C. 创建网络请求接口实例并配置网络请求参数

D. 创建 Observable

2. 在事件处理过程中出异常时，（　　）会被触发，同时队列自动终止，不允许再有事件发出。

A. onNext()　　　　　　　　B. onCompleted()

C. onError()　　　　　　　　D. subscribe()

3. （　　）方法是 RxJava 最基本的创造事件序列的方法。

A. create()　　　　　　　　B. call()

C. subscribe()　　　　　　　D. unsubscribe()

4. 以下哪种不属于 Lambda 表达式写法（　　）。

A. 无参数 + 语句　　　　　　B. 有参数 + 语句

C. 有参数 + 代码块　　　　　D. 无参数 + 代码块

5. 在 Lambda 配置过程中，需要配置以下哪种属于 Lambda 的插件（　　）。

A. apply plugin: 'me.tatarka.retrolambda'

B. compile project('IMLib')

C. compile 'com.squareup.retrofit2:retrofit:2.0.2'

D. compile 'com.squareup.okhttp3:okhttp:3.1.2'

二、填空题

1. RxJava 是一个在_____上使用可观测的序列来组成异步的、基于事件的程序库。

2. _____带来的 Lambda 表达式可以将函数作为参数直接传入方法内。

3. Observable 和 Observer 通过_____方法实现订阅关系，从而 Observable 可以在需要的时候发出事件来通知 Observer。

4. 当 Observable 被订阅的时候，OnSubscribe 的_____方法会自动被调用，事件序列就会依照设定依次触发。

5. 创建了 Observable 和 Observer 之后，再用_____方法将它们联结起来，整条链子就可以工作了。

三、上机题

1. 使用 Rxjava 实现简单的异步点击事件。
2. 使用 Lambda 表达式进行好友删除的编写。

模块三　会话列表

通过会话列表的实现，学习融云 SDK 的基础知识，了解融云 SDK 的使用方法，握融云 IM 界面组件和 IM 通讯能力库，具备通过融云独立开发项目能力。

- 了解融云 SDK 的概念。
- 熟悉融云 SDK 的使用方法。
- 掌握融云 SDK 的插件。
- 具备使用融云开发的能力。

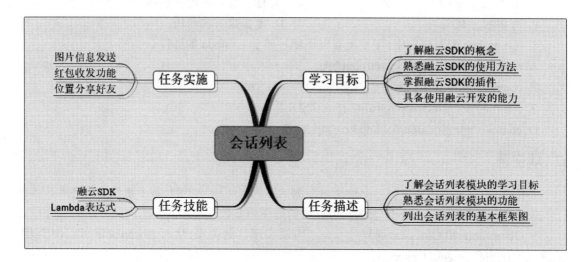

微聊是基于融云 SDK 开发的即时通讯项目,可实现传输视图与音频信息,使用户可以远距离进行直观、真实的音频交流。本模块将通过会话列表的实现,学习融云 SDK 的使用、IMKit 集成的会话界面和 IMLib 提供的基础通讯能力。

【功能描述】

本模块将实现微聊中的会话列表模块。

- 实现发送文字信息和表情功能。
- 实现本地图片发送和拍照发送功能。
- 实现红包收发功能。
- 实现位置分享好友功能。

【基本框架】

基本框架如图 9.1 至图 9.3 所示。

图 9.1　文本聊天示意图

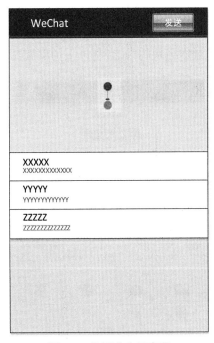

图 9.2　位置分享示意图

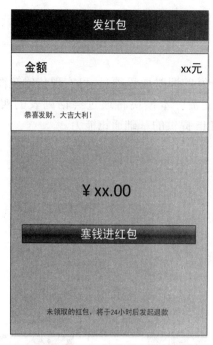

图 9.3 红包发送示意图

通过本模块的学习,将以上的框架图转换成图 9.4 至图 9.8 所示效果。

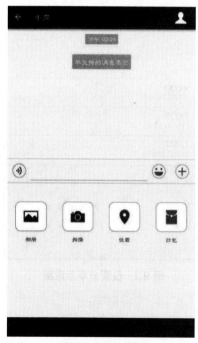

图 9.4 聊天界面

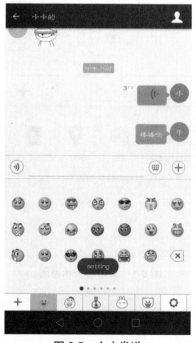

图 9.5 文本发送

项目三 微聊

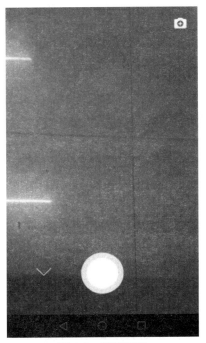

图9.6 拍照界面

图9.7 位置发送

图9.8 红包发送

技能点一 融云 SDK

SDK (Software Development Kit) 是"软体开发工具组",是对一个产品或平台开发应用程式的工具组,由产品的厂商提供给开发者使用。通常是某一家厂商针对某一平台、系统或硬体所发布出来用以开发应用程式的工具组,在这个工具包中,可能包含了各式各样的开发工具,模拟器等。

1 融云 SDK 简介

融云是国内首家专业的即时通讯云服务提供商,为互联网、移动互联网开发者提供免费的即时通讯基础能力和云端服务。通过融云平台,开发者不必搭建服务端硬件环境,就可以将即时通讯、实时网络能力快速集成至应用中。

融云 IM SDK 主要包括:IM 界面组件和 IM 通讯能力库,为方便开发者接入,融云 SDK 还将各部分功能以插件化的形式独立提供,开发者可以根据自己的需要,自由组合下载。各组件的功能如表 9.1 所示。

表 9.1 组件功能

名称	功能介绍
IMKit	集成了会话界面,并且提供了丰富的自定义功能。
IMLib	提供了基础的通信能力,较轻量,适用于对 UI 有较高订制需求的开发者。
CallKit	融云音视频通话的界面组件,包含了单人、多人音视频通话的界面的各种场景和功能。
CallLib	融云音视频通话核心能力组件。
LocationLib	位置相关库文件。
PushLib	融云支持第三方推送(小米),可以从这里下载对应的第三方推送 jar 包。
RedPacket	融云红包相关组件,通过集成该组件,即可快速实现红包功能。

2 融云 SDK 的使用

融云 SDK 的使用方法具体分为以下几个步骤:

1. 注册账号

开发者在集成融云即时通讯、实时网络功能前,需前往融云官方网站注册创建融云开发者帐号。用户注册界面如图 9.9 所示。

图 9.9　用户注册界面

2. 下载 SDK

到融云官方网站下载融云 SDK。融云 SDK 各部分功能以插件化的形式独立提供，开发者可以根据自己的需要，自由组合下载。如图 9.10 所示为融云的分类。

图 9.10　融云 SDK 分类

3. 创建应用

在进行应用开发之前,需要先在融云开发者平台创建应用。界面示意如图 9.11 所示。

图 9.11 创建应用

创建应用后,首先需要了解的是 App Key/Secret,融云 SDK 连接服务器所必须的标识,每一个 App 对应一套 App Key/Secret。针对开发者的生产环境和开发环境,提供两套 App Key/Secret,两套环境的功能完全一致。在应用最终上线前,使用开发环境即可,如图 9.12 所示。

图 9.12 开发环境

4. 获取 Token

Token 的主要作用是身份授权和安全,因此不能通过客户端直接访问融云服务器获取 Token,必须通过 Server API 从融云服务器获取 Token 返回给用户的 App,并在之后连接时使用。

- userId: 每一个用户对应一个 userId,这个 userId 是用户维护的,用户可以直接赋值,两个用户通信,对于融云来说就是两个 userId 间通讯。
- name: 用户的显示名称,用来在 Push 推送时,或者没有传入用户信息时,默认显示的用户名称。
- portraitUri: 用户头像,当没有传入用户信息时作为默认头像,如果图片不存在,IMKit 会显示默认头像。

通过 API 调试,可以得到一个 Token 返回值。就可以直接使用这个 Token 为这位用户进行发送和接受消息。

5. **导入 SDK**

将 SDK 下载完成后对该压缩包减压,如图 9.13 所示:

名称	日期	类型
__MACOSX	2017/12/7 18:25	文件夹
CallKit	2017/12/7 18:25	文件夹
CallLib	2017/12/7 18:25	文件夹
IMKit	2017/12/7 18:25	文件夹
IMLib	2017/12/7 18:25	文件夹
LocationLib	2017/12/7 18:25	文件夹
PushLib	2017/12/7 18:25	文件夹
RedPacket	2017/12/7 18:25	文件夹
release_notes_android.txt	2017/12/7 18:18	文本文档

图 9.13 SDK 压缩包解压

选择需要的组件导入,将 PushLib 中的 jar 包和 pushDaemon → libs 目录下应用所支持平台的 so 拷贝到应用的 libs 目录下,另外还需要将 pushDaemon → executable 目录下各平台的可执行文件 push_daemon 拷贝到应用 Module 的 assets 目录下。如图 9.14 所示:

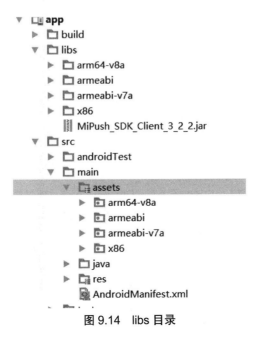

图 9.14 libs 目录

6. 添加配置

打开应用的 build.gradle，在 dependencies 中添加相应模块的依赖。代码如下所示。

```
compile project('IMLib')
```

打开 AndroidManifest.xml 文件，把 meta-data RONG_CLOUD_APP_KEY 的值修改为自己的 AppKey。代码如下所示。

```
<meta-data
    android:name="RONG_CLOUD_APP_KEY"
        android:value=" 应用 AppKey" />
```

在应用的 App Module 的 AndroidManifest.xml 文件中，添加 FileProvider 相关配置，修改 android:authorities 为"应用的包名称 .FileProvider"。代码如下所示。

```
<provider
    android:name="android.support.v4.content.FileProvider"
    android:authorities=" 您的应用包名 .FileProvider"
    android:exported="false"
    android:grantUriPermissions="true">
    <meta-data
        android:name="android.support.FILE_PROVIDER_PATHS"
        android:resource="@xml/rc_file_path" />
</provider>
```

7. 初始化

在整个应用程序全局，只需要调用一次 init 方法。对于快速集成，建议在 App 主进程初始化，只需要使用一句函数，以下为融云 Demo 代码示例：新建一个全局 Application 对象。代码如下所示。

```
public class App extends Application {
    @Override
    public void onCreate() {
        super.onCreate();
        RongIM.init(this);
    }
}
```

8. 连接服务器

连接服务器前，确认已通过融云 Server API 接口获取 Token。如图 9.15 所示：

调试接口	获取 Token	查看对应 Server 开发文档
调试环境	⊙ 开发环境 ○ 生产环境（申请上线通过后才能在生产环境调试 API 接口）
返回数据类型	⊙ json ○ xml
App Key	k51hidwqknbcb	App Secret	**********
userId	App01	用户在 App 中的唯一标识码
name	请输入 name	用户名称
portraitUri	请输入 portraitUri	用户头像 URI

提交

结果：

图 9.15　获取 Token

输入 userId 就可以获取 Token 了，分别输入 10086 和 10010，获取两个 Token，后面用来连接服务器。最后一步，利用获取到的 Token 连接服务器。代码如下所示。

```
/*
连接服务器，在整个应用程序全局，只需要调用一次，需在 {@link #init(Context)}
之后调用，如果调用此接口遇到连接失败，SDK 会自动启动重连机制，进行最多 10 次
重连，分别是 1，2，4，8，16，32，64，128，256，512 秒后。在这之后如果仍没有连接成功，还
会在检测到设备网络状态变化时再次进行重连。
@param token 从服务端获取的用户身份令牌（Token）。
@param callback 连接回调。
@return RongIM 客户端核心类的实例。
*/
private void connect(String token) {
    if (getApplicationInfo().packageName.equals(App.getCurProcessName(getApplication-
Context()))) {
        RongIM.connect(token, new RongIMClient.ConnectCallback() {
/*
Token 错误。可以从下面两点检查：
1. Token 是否过期，如果过期需要向 App Server 重新请求一个新的 Token
2. Token 对应的 appKey 和工程里设置的 appKey 是否一致
*/
```

```java
@Override
public void onTokenIncorrect() {
}
/*
连接融云成功
@param userid 当前 Token 对应的用户 id
*/
@Override public void onSuccess(String userid) {
//userid, 是申请 Token 时填入的 userid
Log.d("LoginActivity", "--onSuccess" + userid);
startActivity(new Intent(LoginActivity.this, MainActivity.class));
finish(); }
/*
连接融云失败
@param errorCode 错误码, 可到官网查看错误码对应的注释
*/
@Override
public void onError(RongIMClient.ErrorCode errorCode) { }   }); } }
```

连接如果成功,会打印出申请 Token 时的 userId,经过测试,正确打印出了 userId,说明连接服务器成功。

9. 配置会话列表

融云 IMKit SDK 使用了 Fragment 作为会话列表和会话界面的组件。下面说明如何在 Activity 里以静态方式加载融云 Fragment。

（1）配置布局文件

会话列表 Activity 对应的布局文件：conversationlist.xml。注意 android:name 固定为融云的 ConversationListFragment。

```xml
<?xml version="1.0" encoding="utf-8"?>
<LinearLayout xmlns:android="http://schemas.android.com/apk/res/android"
android:orientation="vertical"
android:layout_width="match_parent"
android:layout_height="match_parent">
<fragment
android:id="@+id/conversationlist"
    android:name="io.rong.imkit.fragment.ConversationListFragment"
android:layout_width="match_parent"
    android:layout_height="match_parent" />
</LinearLayout>
```

(2) 新建 Activity

```java
public class ConversationListActivity extends FragmentActivity {
@Override
protected void onCreate(Bundle savedInstanceState) {
super.onCreate(savedInstanceState);
setContentView(R.layout.conversationlist);
}
}
```

(3) 配置 intent-filter：

融云 SDK 是通过隐式调用的方式来实现界面跳转的。因此需要在 AndroidManifest.xml 中的会话列表 Activity 下面配置 intent-filter。其中，android:host 是应用的包名，需要手动修改，其他保持不变。

```xml
<!-- 会话列表 -->
<activity
android:name="io.rong.fast.activity.ConversationListActivity"
android:screenOrientation="portrait"
android:windowSoftInputMode="stateHidden|adjustResize">
<intent-filter>
<action android:name="android.intent.action.VIEW" />
<category android:name="android.intent.category.DEFAULT" />
<data
android:host="io.rong.fast"
android:pathPrefix="/conversationlist"
android:scheme="rong" />
</intent-filter>
</activity>
```

10. 启动界面

完成以上配置后，即可启动会话及会话列表界面，启动界面操作必须在执行初始化 SDK 方法以及连接融云服务器 connect 之后进行。代码如下所示：

```
/*
 * @param context 应用上下文。
 * @param conversationType 会话类型。
 * @param targetId 根据不同的 conversationType,可能是用户 Id、讨论组 Id、* 群组 Id 或聊天室 Id。
```

```
 *@param title 聊天的标题，开发者可以在聊天界面通过 *intent.getData().getQuery-
Parameter("title") 获取该值，再手动设置为标题。
 */
public void startConversation(Context context, Conversation.ConversationType conver-
sationType, String targetId, String title)
```

11. 自定义广播接收器

当应用处于后台运行或者连接融云服务器 disconnect() 的时候，如果收到消息，融云 SDK 会以通知形式提醒用户。所以还需要自定义一个继承融云 PushMessageReceiver 的广播接收器，用来接收提醒通知。代码如下所示：

```
public class SealNotificationReceiver extends PushMessageReceiver {
@Override
public boolean onNotificationMessageArrived(Context context, PushNotificationMes-
sage message) {
return false;
// 返回 false, 会弹出融云 SDK 默认通知；返回 true, 融云 SDK 不会弹通知。
}
@Override
public boolean onNotificationMessageClicked(Context context, PushNotificationMes-
sage message) {
return false;
// 返回 false, 会进行融云 SDK 默认处理逻辑，即点击该通知会打开会话列表或
// 会话界面；返回 true, 则由开发者自定义处理逻辑。 } }
```

12. 断开连接

融云 SDK 提供以下两种断开连接的方法。

如果想在断开和服务器的连接后，有新消息时，仍然能够收到推送通知，调用 disconnect() 方法。

```
public void disconnect()
```

如果断开连接后，有新消息时，不想收到任何推送通知，调用 logout() 方法。

```
public void logout()
```

通过以上步骤，即完成了融云 SDK 的集成。接下来就是常见的单聊、群组聊天、讨论组、会话列表等功能的实现，详见各功能模块的实现。

13. API 调用说明

如果基于 IMKit SDK 进行开发，在初始化 SDK 之后，通过 RongIM.getInstance() 方法获取实例，然后调用相应的 api() 方法。代码如下所示：

RongIM.getInstance().setOnReceiveMessageListener(new OnReceiveMessageListener())

拓展：从以上内容可知，SDK 是一个软件开发工具组。在 Android 中还有许多类似的开发工具和工具包。扫描右侧二维码查看常用的 Android 开发工具和工具包。

通过以上技能点的学习，掌握如何集成融云 SDK，以下将学习如何使用 SDK 实现会话列表功能。

第一步：会话置顶，取消置顶。具体代码如 CORE0901 所示。

代码 CORE0901　会话置顶，取消置顶

```
private void setAdapter() {
    // 长按适配器触发长按事件
    mAdapter.setOnItemLongClickListener((helper, parent, itemView, position) -> {
        // 长按的标签与 list 中的标签位置对应
        Conversation item = mData.get(position);
        // 自定义 View 组件
        View conversationMenuView = View.inflate(
                mContext, R.layout.dialog_conversation_menu,null);
        // 弹出选择弹窗
        mConversationMenuDialog = new CustomDialog(
                mContext, conversationMenuView, R.style.MyDialog);
        TextView tvSetConversationToTop = (TextView)
                conversationMenuView.findViewById(R.id.tvSetConversationToTop);
        // 判断是置顶消息操作还是取消置顶的操作
        tvSetConversationToTop.setText(item.isTop() ?
                UIUtils.getString(R.string.cancel_conversation_to_top) :
                UIUtils.getString(R.string.set_conversation_to_top));
        conversationMenuView.findViewById(R.id.tvSetConversationToTop).
                setOnClickListener(v ->
                // 请求数据，需要置顶消息
                RongIMClient.getInstance().setConversationToTop(
                        item.getConversationType(), item.getTargetId(),
                        !item.isTop(), new RongIMClient.ResultCallback<Boolean>() {
                    // 响应请求
```

```java
            @Override
            public void onSuccess(Boolean aBoolean) {
                loadData();        // 响应解析并填入 List
                mConversationMenuDialog.dismiss();
                mConversationMenuDialog = null;
            }
            @Override
            public void onError(RongIMClient.ErrorCode errorCode) {
            }
        }));
    // 请求数据,取消置顶
    conversationMenuView.findViewById(R.id.tvDeleteConversation).setOnClickListener(v -> {
        RongIMClient.getInstance().removeConversation(
            item.getConversationType(), item.getTargetId(),
            new RongIMClient.ResultCallback<Boolean>() {
    // 请求响应
    @Override
    public void onSuccess(Boolean aBoolean) {
        loadData();        // 响应解析并填入 List
        mConversationMenuDialog.dismiss();
        mConversationMenuDialog = null;
    }
    // 响应异常
            @Override
            public void onError(RongIMClient.ErrorCode errorCode) {
            } }); });
        mConversationMenuDialog.show();     // 显示弹窗
            return true;
        });
        // 重新加载适配器
        getView().getRvRecentMessage().setAdapter(mAdapter);
    }}
    // 解析响应并填充方法
    private void loadData() {
        RongIMClient.getInstance().getConversationList(
            new RongIMClient.ResultCallback<List<
```

```
Conversation>>() {
    @Override
    public void onSuccess(List<Conversation> conversations) {
        if (conversations != null && conversations.size() > 0) {
            mData.clear();
            mData.addAll(conversations);
            filterData(mData);
        }
    }
    @Override
    public void onError(RongIMClient.ErrorCode errorCode) {
        LogUtils.e(" 加载最近会话失败：" + errorCode);
    }
});}
```

第二步：删除会话或者撤回消息。具体代码如 CORE0902 所示。

代码 CORE0902　删除会话，撤回消息

```
public void setAdapter() {
if (mAdapter == null) {
// 长按触发事件响应
mAdapter.setOnItemLongClickListener((helper, viewGroup, view, position) -> {
View sessionMenuView = View.inflate(
    mContext, R.layout.dialog_session_menu, null);
    mSessionMenuDialog = new CustomDialog(
mContext, sessionMenuView, R.style.MyDialog);
// 撤回消息文本
TextView tvReCall = (TextView) sessionMenuView.findViewById(R.id.tvReCall);
// 删除消息文本
TextView tvDelete = (TextView) sessionMenuView.findViewById(R.id.tvDelete);
// 根据消息类型控制显隐
Message message = mData.get(position);
MessageContent content = message.getContent();
if (content instanceof GroupNotificationMessage ||
content instanceof RecallNotificationMessage) {
    return false;
}
if (content instanceof RedPacketMessage ||
!message.getSenderUserId().equalsIgnoreCase(UserCache.getId())) {
    tvReCall.setVisibility(View.GONE);
```

```java
}
// 点击撤回消息文本，请求数据
tvReCall.setOnClickListener(v ->
RongIMClient.getInstance().recallMessage(message, "",
        new RongIMClient.ResultCallback<RecallNotificationMessage>() {
// 回传响应
@Override
public void onSuccess(RecallNotificationMessage recallNotificationMessage) {
    UIUtils.postTaskSafely(() -> {
        // 撤回成功
        recallMessageAndInsertMessage(recallNotificationMessage, position);
        mSessionMenuDialog.dismiss();
        mSessionMenuDialog = null;
        UIUtils.showToast(UIUtils.getString(R.string.recall_success));
    }); }
@Override
public void onError(RongIMClient.ErrorCode errorCode) {
    UIUtils.postTaskSafely(() -> {
        // 撤回失败
        mSessionMenuDialog.dismiss();
        mSessionMenuDialog = null;
        UIUtils.showToast(UIUtils.getString(R.string.recall_fail) + ":"
                + errorCode.getValue());
    }); }}));
    // 点击删除消息文本，请求数据
    tvDelete.setOnClickListener(v ->
  RongIMClient.getInstance().deleteMessages(new
  int[]{message.getMessageId()},
  new RongIMClient.ResultCallback<Boolean>() {
@Override
// 回传响应
public void onSuccess(Boolean aBoolean) {
    UIUtils.postTaskSafely(() -> {
        // 删除成功
        mSessionMenuDialog.dismiss();
        mSessionMenuDialog = null;
        mData.remove(position);
```

```
                mAdapter.notifyDataSetChangedWrapper();
                UIUtils.showToast(UIUtils.getString(R.string.delete_success));
        }); }
        @Override
        public void onError(RongIMClient.ErrorCode errorCode) {
            UIUtils.postTaskSafely(() -> {
                // 删除失败
                mSessionMenuDialog.dismiss();
                mSessionMenuDialog = null;
                UIUtils.showToast(UIUtils.getString(R.string.delete_fail) + ":" +
                errorCode.getValue());
            }); }}));
            mSessionMenuDialog.show();                    // 弹窗显示
                return false;
        });}
    else {
            mAdapter.notifyDataSetChangedWrapper();       // 数据全局刷新
    }}
```

第三步:实现发送文本消息的功能。具体代码如 CORE0903 所示。

代码 CORE0903　发送文本消息

```
public void sendTextMsg() {
sendTextMsg(getView().getEtContent().getText().toString());getView().
getEtContent().setText("");
}
public void sendTextMsg(String content) {
    RongIMClient.getInstance().sendMessage(mConversationType, mSessionId,
            TextMessage.obtain(content), mPushCotent, mPushData,
            new RongIMClient.SendMessageCallback() {// 发送消息的回调
                @Override
                public void onError(Integer integer,
                RongIMClient.ErrorCode errorCode) {
                    updateMessageStatus(integer);
                }
                @Override
                public void onSuccess(Integer integer) {
                    updateMessageStatus(integer);
```

```
                }
            }, new RongIMClient.ResultCallback<Message>() {
                // 消息存库的回调,可用于获取消息实体
                @Override
                public void onSuccess(Message message) {
                    mAdapter.addLastItem(message);
                    rvMoveToBottom();
                }
                @Override
                public void onError(RongIMClient.ErrorCode errorCode) {

                }
            });
}
```

通过以上代码实现如图 9.16 效果。

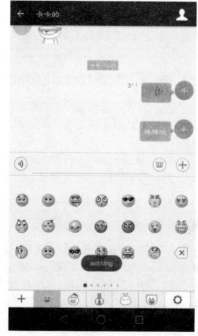

图 9.16 文本发送

第四步:发送图片消息。具体代码如 CORE0904 所示。

代码 CORE0904 发送图片消息

```
public void sendImgMsg(Uri imageFileThumbUri, Uri imageFileSourceUri) {
    ImageMessage imgMsg = ImageMessage.obtain(imageFileThumbUri,
```

```java
            imageFileSourceUri);
        RongIMClient.getInstance().sendImageMessage(mConversationType,
        mSessionId, imgMsg,
mPushCotent, mPushData,
            new RongIMClient.SendImageMessageCallback() {
                @Override
                public void onAttached(Message message) {
                    // 保存数据库成功
                    mAdapter.addLastItem(message);
                    rvMoveToBottom();
                }
                @Override
                public void onError(Message message,
                    RongIMClient.ErrorCode errorCode) {
                    // 发送失败
                    updateMessageStatus(message);
                }
                @Override
                public void onSuccess(Message message) {
                    // 发送成功
                    updateMessageStatus(message);
                }
                @Override
                public void onProgress(Message message, int progress) {
                    // 发送进度
                    message.setExtra(progress + "");
                    updateMessageStatus(message);
                }           });}
public void sendImgMsg(File imageFileThumb, File imageFileSource) {
        Uri imageFileThumbUri = Uri.fromFile(imageFileThumb);
        Uri imageFileSourceUri = Uri.fromFile(imageFileSource);
        sendImgMsg(imageFileThumbUri, imageFileSourceUri);
    }
```

通过以上代码实现效果如图 9.17 所示。

图 9.17　图片选择

第五步：发送贴图消息。具体代码如 CORE0905 所示。

代码 CORE0905　发送贴图消息

```
public void sendFileMsg(File file) {
Message fileMessage = Message.obtain(mSessionId, mConversationType,
FileMessage.obtain(
Uri.fromFile(file)));
RongIMClient.getInstance().sendMediaMessage(fileMessage, mPushCotent,
mPushData,
new IRongCallback.ISendMediaMessageCallback() {
    @Override
    public void onProgress(Message message, int progress) {
        // 发送进度
        message.setExtra(progress + "");
        updateMessageStatus(message);
    }
    @Override
    public void onCanceled(Message message) {
    }
    @Override
```

```java
    public void onAttached(Message message) {
        // 保存数据库成功
        mAdapter.addLastItem(message);
        rvMoveToBottom();
    }
    @Override
    public void onSuccess(Message message) {
        // 发送成功
        updateMessageStatus(message);
    }
    @Override
    public void onError(Message message, RongIMClient.ErrorCode errorCode) {
        // 发送失败
        updateMessageStatus(message);
    }    });}
```

第六步:实现发送当前位置信息的功能。具体代码如 CORE0906 所示。

代码 CORE0906　发送位置消息

```java
public void sendLocationMessage(LocationData locationData) {
    LocationMessage message = LocationMessage.obtain(
    locationData.getLat(), locationData.getLng(),
    locationData.getPoi(), Uri.parse(locationData.getImgUrl()));
    RongIMClient.getInstance().sendLocationMessage(
    Message.obtain(mSessionId, mConversationType,message),
    mPushCotent, mPushData, new IRongCallback.ISendMessageCallback() {
        @Override
        public void onAttached(Message message) {
            // 保存数据库成功
            mAdapter.addLastItem(message);
            rvMoveToBottom();
        }
        @Override
        public void onSuccess(Message message) {
            // 发送成功
            updateMessageStatus(message);
        }
        @Override
```

```
public void onError(Message message, RongIMClient.ErrorCode errorCode) {
    // 发送失败
    updateMessageStatus(message);
}
});}
```

通过以上代码实现如图 9.18 所示效果。

图 9.18　位置发送

第七步：发送语音消息。具体代码如 CORE0907 所示。

代码 CORE0907　发送语音消息

```
public void sendAudioFile(Uri audioPath, int duration) {
if (audioPath != null) {
    // 得到语音路径
    File file = new File(audioPath.getPath());
    if (!file.exists() || file.length() == 0L) {
        LogUtils.sf(UIUtils.getString(R.string.send_audio_fail));
        return;
    }
    // 语音消息
```

```
            VoiceMessage voiceMessage = VoiceMessage.obtain(audioPath, duration);
            // 网络请求，发送语音消息
            RongIMClient.getInstance().sendMessage(Message.obtain(mSessionId,
            mConversationType, voiceMessage), mPushCotent, mPushData,
            new IRongCallback.ISendMessageCallback() {
                @Override
                public void onAttached(Message message) {
                    // 保存数据库成功
                    mAdapter.addLastItem(message);
                    rvMoveToBottom();
                }
                @Override
                public void onSuccess(Message message) {
                    // 发送成功
                    updateMessageStatus(message);
                }
                @Override
            public void onError(Message message, RongIMClient.ErrorCode errorCode) {
                    // 发送失败
                    updateMessageStatus(message);
                }
            });}
```

第八步：实现发送红包的功能。具体代码如 CORE0908 所示。

代码 CORE0908　　发红包
| ```
public void sendRedPacketMsg() {
if (mConversationType == Conversation.ConversationType.PRIVATE) {
    // 获取接收人信息
    UserInfo userInfo = DBManager.getInstance().getUserInfo(mSessionId);
    if (userInfo != null)
    // 发送红包请求
    RedPacketUtil.startRedPacket(mContext, userInfo, RPSendPacketCallback);
} else {
    // 获取群组信息
    List<GroupMember> groupMembers = DBManager.getInstance().getGroupMembers(mSessionId);
    if (groupMembers != null)
``` |
```

```
        RedPacketUtil.startRedPacket(mContext, mSessionId, groupMembers.size(),
         RPSendPacketCallback);
        }}
        // 发送红包信息后返回数据
        RPSendPacketCallback RPSendPacketCallback = new RPSendPacketCallback(){
              @Override
              public void onGenerateRedPacketId(String redPacketId) {
              }
              @Override
              public void onSendPacketSuccess(RedPacketInfo redPacketInfo) {
// 发送红包信息
RedPacketMessage rpMsg = RedPacketMessage.obtain(redPacketInfo.redPacketId,
    redPacketInfo.fromNickName, redPacketInfo.redPacketType,
    redPacketInfo.redPacketGreeting);
// 请求数据
RongIMClient.getInstance().sendMessage(Message.obtain(mSessionId,
    mConversationType, rpMsg), mPushCotent, mPushData,
    new IRongCallback.ISendMessageCallback() {
              @Override
              public void onAttached(Message message) {
                  // 保存数据库成功
                  mAdapter.addLastItem(message);
                  rvMoveToBottom();
              }
              @Override
              public void onSuccess(Message message) {
                  // 发送成功
                  updateMessageStatus(message);
              }
              @Override
              public void onError(Message message,
              RongIMClient.ErrorCode errorCode) {
                  // 发送失败
                  updateMessageStatus(message);
              }      });    }};
```

通过以上代码实现效果如图 9.19 所示。

图 9.19 红包发送

本模块介绍了微聊会话列表的实现,通过本模块的学习可以了解融云 SDK 的基础知识,掌握 SDK 插件的使用方法,学习之后能够实现聊天、发送图片、发红包、发送地理位置、发送贴图信息等功能。

对 SDK 有了基本的了解之后学习如何从零开始一步一步接入 SDK,每个渠道或者服务商面对开发者都会提供相应的 SDK,里面包含相应的开发文档,包括开发 Demo 还有 jar 包或者项目所需的资源。此外产品厂商还有不同的 SDK 提供给开发者使用,就是接下来介绍的 ShareSDK。

技能扩展——ShareSDK

1 简介

ShareSDK 是一种社会化分享组件，为 iOS、Android、WP8 的 APP 提供社会化功能，集成了一些常用的类库和接口，缩短开发者的开发时间，还有社会化统计分析管理后台。支持许多主流社交平台，帮助开发者轻松实现社会化分享、登录、关注、获得用户资料、获取好友列表等主流的社会化功能，强大的统计分析管理后台，可以实时了解用户、信息流、回流率、传播效率等数据。

2 ShareSDK 的配置

以下是 ShareSDK 配置的具体步骤。

第一步：百度搜索"mob"，找到 Mob 官网：http://www.mob.com/。如图 9.20 所示。

图 9.20 Mob 官网

第二步：点击 SDK 下载，加载图 9.21 界面。

项目三 微聊

图 9.21 点击安卓图标

第三步：点击安卓图像，进入下载页。如图 9.22 所示。

图 9.22 点击 SDK 下载

第四步：点击立即下载，将要用到的 jar 包进行选中进行下载。如图 9.23 所示。

图 9.23 下载 SDK

第五步:点击下载,指定路径下载至本地计算机。将下载的压缩文件进行解压。如图 9.24 所示。

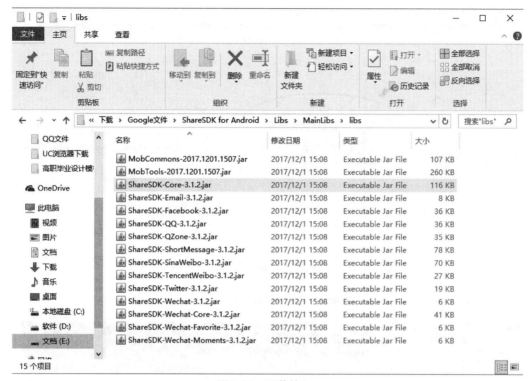

图 9.24 下载的 jar

第六步:将所有 jar 包复制进入项目中,再进行添加。

3 配置 AndroidManifest.xml

添加如下权限:(和 application 节点并列)。

```
<uses-permission android:name="android.permission.GET_TASKS"/>
<uses-permission android:name="android.permission.INTERNET"/>
<uses-permission android:name="android.permission.ACCESS_WIFI_STATE"/>
<uses-permission android:name="android.permission.ACCESS_NETWORK_STATE"/>
<uses-permission android:name="android.permission.CHANGE_WIFI_STATE"/>
<uses-permission android:name="android.permission.WRITE_EXTERNAL_STORAGE"/>
<uses-permission android:name="android.permission.READ_PHONE_STATE"/>
<uses-permission android:name="android.permission.MANAGE_ACCOUNTS"/>
<uses-permission android:name="android.permission.GET_ACCOUNTS"/>
```

在 application 节点下注册下面的 Activity。

```xml
<activity
        android:name="com.mob.tools.MobUIShell"
        android:theme="@android:style/Theme.Translucent.NoTitleBar"
        android:configChanges="keyboardHidden|orientation|screenSize"
        android:screenOrientation="portrait"
        android:windowSoftInputMode="stateHidden|adjustResize" >
    <intent-filter>
    <data android:scheme="tencent1104646053" />
    <action android:name="android.intent.action.VIEW" />
    <category android:name="android.intent.category.BROWSABLE" />
    <category android:name="android.intent.category.DEFAULT" />
    </intent-filter>
</activity>
```

同时,在清单文件中进行声明。

```xml
    <!-- 微信分享回调 -->
<activity
    android:name=".wxapi.WXEntryActivity"
    android:configChanges="keyboardHidden|orientation|screenSize"
    android:exported="true"
    android:screenOrientation="portrait"
    android:theme="@android:style/Theme.Translucent.NoTitleBar"/>
<!-- 支付宝分享回调 -->
<activity
    android:name=".apshare.ShareEntryActivity"
    android:configChanges="keyboardHidden|orientation|screenSize"
    android:exported="true"
    android:theme="@android:style/Theme.Translucent.NoTitleBar"/>
```

Message	通讯	Plug-in	插件
Push	推送	Subassembly	组件
Integration	集成	Open source	开源
Call	调用	Acceptor	接收器
Location	位置	Mechanism	机制

一、选择题

1. 下列（   ）是快速实现红包功能。
   A. RedPacket   B. PushLib   C. LocationLib   D. CallKit
2. 如果在断开和融云的连接后，有新消息时，仍然能够收到推送通知，调用（   ）方法。
   A. disconnect()   B. logout()   C. setOnClickListener()   D. onClick()
3. 如果断开连接后，有新消息时，不想收到任何推送通知，调用（   ）方法。
   A. disconnect()   B. logout()   C. setOnClickListener()   D. onClick()
4. ShareSDK 是一种（   ）组件，为 iOS、Android、WP8 的 APP 提供社会化功能，集成了一些常用的类库和接口。
   A. 分享   B. 会话   C. 音频   D. 缓存
5. 融云 IMKit SDK 使用了（   ）作为会话列表和会话界面的组件。
   A. Layout   B. Fragment   C. View   D. Toolbar

二、填空题

1. 融云 IM SDK 主要包括_____和 IM 通讯能力库。
2. Token 的主要作用是_____和_____，因此不能通过客户端直接访问融云服务器获取 Token，必须通过 Server API 从融云服务器获取 Token 返回给您的 App，并在之后连接时使用。
3. 融云 SDK 各部分功能以_____的形式独立提供，开发者可以根据自己的需要，自由组合下载。
4. _____和_____是融云 SDK 连接服务器所必须的标识。
5. ShareSDK 是一种_____组件，为 iOS、Android、WP8 的 APP 提供社会化功能，集成了一些常用的类库和接口。

三、上机题

1. 完成融云 SDK 的集成。
2. 编写代码实现会话列表。